A Tutorial Introduction to VHDL Programming

Orhan Gazi

A Tutorial Introduction to VHDL Programming

 Springer

Orhan Gazi
Department of Electronics and
  Communication Engineering
Çankaya University
Ankara, Turkey

ISBN 978-981-13-4764-1   ISBN 978-981-13-2309-6   (eBook)
https://doi.org/10.1007/978-981-13-2309-6

This Springer imprint is published by the registered company Springer Nature Singapore Pte Ltd.
The registered company address is: 152 Beach Road, #21-01/04 Gateway East, Singapore 189721,
Singapore

Contents

Chapter 1
Entity, Architecture and VHDL Operators

VHDL stands for "very high speed integrated circuits hardware description language", and first developed by the US department of defense in 1980s. It is used to describe the behavior of an electronic circuit. VHDL programs consist of program modules. A module is nothing but a program segment written for some specific purposes. For instance, we have package modules, test modules, and main VHDL modules, i.e., main programs.

In this chapter, entity and architecture parts of a VHDL program will be explained in details. Entity part is used to describe the input and output ports of an electronic circuit, i.e., used to describe I/O ports. On the other hand, architecture part of a VHDL module is used to describe the operation of the electronic circuit. During the circuit operation, the circuit process its input data obtained from input ports, and produces its output data to be sent to the output ports. We will also discuss the data types used for the **ports** available in **entity** part. In addition, VHDL operators will be explained shortly.

1.1 Entity

As we mentioned in the previous paragraph, **entity** part of the VHDL program describes the I/O ports of an electronic circuit. A port can be an input port, or an output port, or both input and output port at the same time, or can be a buffer port. In Fig. 1.1, the graphical illustration of the port types is given.

© Springer Nature Singapore Pte Ltd. 2019
O. Gazi, *A Tutorial Introduction to VHDL Programming*,
https://doi.org/10.1007/978-981-13-2309-6_1

Fig. 1.1 Black box
representation of an electronic
circuit

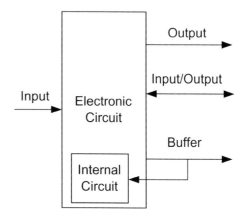

The general structure of the **entity** part is shown in PR 1.1

PR 1.1 General structure of
entity unit

entity entity_name **is**
  port(port_ID1: I/O option signal_type;
        port_ID2: I/O option signal_type;
        ···**);**
end [entity] [entity_name];

where '[entity] [entity_name]' means that you can omit or keep words inside
braces. Let's illustrate the use of entity by an example.

Example 1.1 The black box representation of an electronic circuit is shown in
Fig. 1.2. Describe the electronic circuit ports by a VHDL program, i.e., by PR.

Fig. 1.2 Black box
representation of an electronic
circuit for Example 1.1

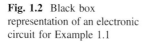

Solution 1.1 We can write the **entity** part for the black box shown in Fig. 1.2 with
the following steps.

(S1) First, write the reserved word entity as in PR 1.2.

PR 1.2 Program 1.2

entity

(S2) In step 2, we give a name to entity as in PR 1.3.

PR 1.3 Program 1.3

entity my_circuit_name

(S3) In step 3, write the reserved word **is** to the end of the line of PR 1.3 as in PR 1.4.

PR 1.4 Program 1.4

```
entity my_circuit_name is
```

(S4) In step 4, write the closing tag of entity part as in PR 1.5.

PR 1.5 Program 1.6

```
entity my_circuit_name is

end my_circuit_name
```

(S5) In step 5, put a semicolon to the end of the closing tag as in PR 1.6.

PR 1.6 Program 1.6

```
entity my_circuit_name is

end my_circuit_name;
```

(S6) In step 6, write the reserved word **port** as in PR 1.7.

PR 1.7 Program 1.7

```
entity my_circuit_name is
   port
end my_circuit_name;
```

(S7) In step 7, add parentheses to the **port** as in PR 1.8.

PR 1.8 Program 1.8

```
entity my_circuit_name is
   port()
end my_circuit_name;
```

(S8) In step 8, put a semicolon to the end of **port**() as in PR 1.9.

PR 1.9 Program 1.9

```
entity my_circuit_name is
   port();
end my_circuit_name;
```

(S9) If we look at the black box in Fig. 1.2, we see that we have two input ports and three output ports. Let's give names to these ports as in PR 1.10.

PR 1.10 Program 1.10

```
entity my_circuit_name is
   port( inp1
         inp2
         outp1
         outp2
         outp3 );
end my_circuit_name;
```

(S10) In step 10, we indicate the type of the ports using the reserved words **in** and **out** as in PR 1.11.

PR 1.11 Program 1.11

```
entity my_circuit_name is
   port( inp1: in
         inp2: in
         outp1: out
         outp2: out
         outp3: out );
end my_circuit_name;
```

(S11) In Step 11, we indicate the data type available at the input and output ports an in PR 1.12 where we used **std_logic** for all port data types.

PR 1.12 Program 1.12

```
entity my_circuit_name is
   port( inp1: in  std_logic
         inp2: in  std_logic
         outp1: out  std_logic
         outp2: out  std_logic
         outp3: out  std_logic );
end my  circuit  name;
```

(S12) In step 12, we put semicolon to the end of every **std_logic** except for the last one as in PR 1.13.

PR 1.13 Program 1.13

```
entity my_circuit_name is
   port( inp1: in  std_logic;
         inp2: in  std_logic;
         outp1: out  std_logic;
         outp2: out  std_logic;
         outp3: out  std_logic );
end my_circuit_name;
```

The enclosing tag or enclosing line of the program segment in PR 1.13 can also be written as in PR 1.14. We can also write the enclosing line as "**end entity** my_circuit_name".

```
entity my_circuit_name is
   port( inp1: in  std_logic;
         inp2: in  std_logic;
         outp1: out  std_logic;
         outp2: out  std_logic;
         outp3: out  std_logic );
end entity;
```

```
entity my_circuit_name is
   port( inp1: in  std_logic;
         inp2: in  std_logic;
         outp1: out  std_logic;
         outp2: out  std_logic;
         outp3: out  std_logic );
end;
```

PR 1.14 Program 1.14

It is up to the programmer to select the enclosing line of the **entity** part.

(S12) In Step 12, add the header lines which are necessary for the data type **std_logic** to be meaningful in our VHDL program as in PR 1.15.

PR 1.15   Program 1.15

```
library IEEE;
use IEEE.std_logic_1164.all;

entity my_circuit_name is
  port( inp1: in  std_logic;
        inp2: in  std_logic;
        outp1: out  std_logic;
        outp2: out  std_logic;
        outp3: out  std_logic );;
end my  circuit  name;
```

(S12) Finally, we can add some comments to our VHDL code. For this purpose, we can use

−− **Your comment here**

format in our program. Then, our **entity** part becomes as in PR 1.16.

```
library IEEE; -- IEEE library
use IEEE.std_logic_1164.all;  -- Necessary to use the std_logic

entity my_circuit_name is
  port( inp1: in  std_logic; --Input port
        inp2: in  std_logic; --Input port
        outp1: out  std_logic; --Output port
        outp2: out  std_logic; --Output port
        outp3: out  std_logic ); --Output port
end my_circuit_name; -- End of entity
```

PR 1.16   Program 1.16

The VHDL program in PR 1.16 completely describes the input/output ports of the electronic circuit whose black box representation is given in Fig. 1.2.

The VHDL program in PR 1.16 can also be written as in PR 1.17 where the port types are written in the same line.

```
library IEEE; -- IEEE library
use IEEE.std_logic_1164.all;  -- Necessary to use the std_logic

entity my_circuit_name is
  port( inp1, inp2: in  std_logic; --Input ports
        outp1, outp2, outp3: out  std_logic ); --Output ports
end my_circuit_name; -- End of entity
```

PR 1.17   Program 1.17

1.2   The Data Types Used in Input/Output Ports

In this section, we will give information about the data types used in VHDL programming.

std_logic

In PR 1.17 the data type used for the ports is **std_logic** defined in the package IEEE. std_logic_1164.all. For **std_logic** data type, there are 8 possible values available, and these values are tabulated in Table 1.1.

Table 1.1 std_logic values	
	'X' Unknown
	'0' Logic 0
	'1' Logic 1
	'Z' High Impedance
	'W' Weak Unknown
	'L' Weak Low
	'H' Weak High
	'-' Don't Care

std_logic_vector

The **std_logic_vector** data type is defined in the library IEEE.std_logic_1164.all. If an I/O port has data type of std_logic_vector, it means that the I/O port has a number of **std_logic** values.

std_ulogic

The data type **std_ulogic** is defined in the package IEEE.std_logic_1164.all. For **std_ulogic** data type, there are 9 possible values available, and these values are tabulated in Table 1.2.

Table 1.2 std_ulogic values	
	'U' Uninitialized
	'X' Unknown
	'0' Logic 0
	'1' Logic 1
	'Z' High Impedance
	'W' Weak Unknown
	'L' Weak Low
	'H' Weak High
	'-' Don't Care

std_ulogic_vector

The **std_ulogic_vector** data type is defined in the library IEEE.std_logic_1164.all. If an I/O port has data type of std_ulogic_vector, it means that the I/O port has a number of **std_ulogic** values.

bit

The **bit** data type is defined in standard package, i.e., we don't need to include an extra package at the header of the VHDL program. For instance, if we use data type **bit** for port I/Os, we can write PR 1.17 as in PR 1.18 where it is seen that no header file is needed.

PR 1.18 Program 1.18

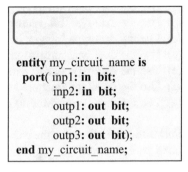

```
entity my_circuit_name is
  port( inp1: in bit;
        inp2: in bit;
        outp1: out bit;
        outp2: out bit;
        outp3: out bit);
end my_circuit_name;
```

bit_vector

The **bit_vector** data type is defined in standard package, i.e., we don't need to include an extra package at the header of the VHDL program. If an I/O port has data type of **bit_vector**, it means that the I/O port has a number of **bit** values.

integer, natural, positive

The data types **integer, natural, positive** are defined in standard package, i.e., we don't need to include an extra package at the header of the VHDL program. The data type **integer** is used to represent integer numbers in the range -2^{31} to $2^{31} - 1$. The data type **natural** is used to represent the integers in the range 0 to $2^{31} - 1$. On the other hand, the data type **positive** is used to represent the integers in the range 1 to $2^{31} - 1$.

unsigned, signed

The data types **unsigned** and **signed** are defined in the packages **numeric_std** and **std_logic_arith**. One of these packages should be included in the header of the VHDL program to be able to use these data types in **entity** declarations. The data type **unsigned** is used to represent unsigned numbers, i.e., nonnegative integers, and the data type **signed** is used to represent the **signed** numbers, i.e., integers.

An electronic circuit with a number of ports which consists of a number of bits can be expressed with a number of equivalent ways. Let's illustrate this concept with an example.

Example 1.2 The black box representation of an electronic circuit is shown in Fig. 1.3. Describe the electronic circuit ports by a VHDL program.

Fig. 1.3 The black box
representation of an electronic
circuit for Example 1.2

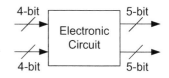

(a) Use the data type **std_logic_vector** to describe the ports in **entity** part.
(b) Use the data type **bit_vector** to describe the ports in **entity** part.
(c) Use the data type **integer** to describe the ports in **entity** part.
(d) Use the data type **natural** to describe the ports in **entity** part.
(e) Use the data type **positive** to describe the ports in **entity** part.
(f) Use the data type **unsigned** to describe the ports in **entity** part.
(g) Use the data type **signed** to describe the ports in **entity** part.

Solution 1.2 Let's solve the part-a in details, and follow the same approach for the
other parts in short.

In Fig. 1.3, the number of bits available at the ports is indicated, however, the
type of the data available at the ports is not clarified. For this reason, we can employ
all the data types mentioned in the question to represent the bits. Otherwise, considering the type of the data available at the ports, we would select the suitable
VHDL data type representation in our VHDL program. Now let's proceed with the
solution of part-a.

(a) We can write the **entity** part for the black box as shown in Fig. 1.3 with the
    following steps.

(S1) First, write the reserved word entity as in PR 1.19.

PR 1.19 Program 1.19

> entity

(S2) In step 2, we give a name to entity as in PR 1.20.

PR 1.20 Program 1.20

> **entity** FourBit_Circuit

(S3) In step 3, write the reserved word **is** to the end of the line of PR 1.20 as in PR
1.21.

PR 1.21 Program 1.21

> **entity** FourBit_Circuit **is**

(S4) In step 4, write the closing tag of entity part as in PR 1.22.

PR 1.22 Program 1.22

> **entity** FourBit_Circuit **is**
>
> **end** FourBit_Circuit

(S5) In step 5, put a semicolon to the end of the closing tag as in PR 1.23.

PR 1.23 Program 1.23

```
entity FourBit_Circuit is
                                    '
end FourBit_Circuit;
```

(S6) In step 6, write the reserved word **port** as in PR 1.24.

PR 1.24 Program 1.24

```
entity FourBit_Circuit is
port
end FourBit_Circuit;
```

(S7) In step 7, add parentheses to the **port** as in PR 1.25.

PR 1.25 Program 1.25

```
entity FourBit_Circuit is
port()
end FourBit_Circuit;
```

(S8) In step 8, put a semicolon to the end of **port()** as in PR 1.26.

PR 1.26 Program 1.26

```
entity FourBit_Circuit is
port();
end FourBit_Circuit;
```

(S9) If we look at the black box in Fig. 1.3, we see that we have two input ports and two output ports. Let's give names to these ports as in PR 1.27.

PR 1.27 Program 1.27

```
entity FourBit_Circuit is
port( inp1
      inp2
      outp1
      outp2);
end FourBit_Circuit;
```

(S10) In step 10, we indicate the type of the ports using the reserved words **in** and **out** as in PR 1.28.

PR 1.28 Program 1.28

```
entity FourBit_Circuit is
port( inp1: in
      inp2: in
      outp1: out
      outp2: out );
end FourBit_Circuit;
```

(S11) In Step 11, we indicate the data type available at the input and output ports an in PR 1.29 where we used **std_logic_vector** for all port data types.

PR 1.29 Program 1.29

```
entity FourBit_Circuit is
  port( inp1: in  std_logic_vector(3 downto 0)
        inp2: in  std_logic_vector(3 downto 0)
        outp1: out  std_logic_vector(4 downto 0)
        outp2: out  std_logic_vector(4 downto 0) );
end FourBit_Circuit;
```

(S12) In step 12, we put semicolon to the end of every **std_logic_vector** except for the last one as in as in PR 1.30.

PR 1.30 Program 1.30

```
entity FourBit_Circuit is
  port( inp1: in  std_logic_vector(3 downto 0);
        inp2: in  std_logic_vector(3 downto 0);
        outp1: out  std_logic_vector(4 downto 0);
        outp2: out  std_logic_vector(4 downto 0) );
end FourBit_Circuit;
```

The enclosing tag or enclosing line of the program segment in PR 1.30 can also be written as in PR 1.31.

PR 1.31 Program 1.31

```
entity FourBit_Circuit is
  port( inp1: in  std_logic_vector(3 downto 0);
        inp2: in  std_logic_vector(3 downto 0);
        outp1: out  std_logic_vector(4 downto 0);
        outp2: out  std_logic_vector(4 downto 0) );
end entity;
```

Another alternative for the closing tag of the entity is to use "**end;**" alone at the end of the declaration. Besides, we can also use "**end entity** FourBit_Circuit" fort he enclosing lines of the entity unit. It is up to the programmer to select the enclosing line of the **entity** part.

(S13) In Step 12, add the header lines which are necessary for the data type **std_logic_vector** to be meaningful in our VHDL program as in PR 1.32.

PR 1.32 Program 1.32

```
library IEEE;
use IEEE.std_logic_1164.all;

entity FourBit_Circuit is
  port( inp1: in  std_logic_vector(3 downto 0);
        inp2: in  std_logic_vector(3 downto 0);
        outp1: out  std_logic_vector(4 downto 0);
        outp2: out  std_logic_vector(4 downto 0) );
end entity;
```

(b) We can write the **entity** segment using the data type **bit_vector**. However, in this case since **bit_vector** is defined in standard package, we don't need to include any header file. Then, **entity** segment can be written as in **PR** 1.33.

PR 1.33 Program 1.33

```
entity FourBit_Circuit is
  port( inp1: in  bit_vector(3 downto 0);
        inp2: in  bit_vector(3 downto 0);
        outp1: out  bit_vector(4 downto 0);
        outp2: out  bit_vector(4 downto 0) );
end entity;
```

(c) We can write the **entity** segment using the data type **integer**. The data type **integer is** defined in standard package; we don't need to include any header file. However, it is logical to indicate a range for the **integer** data type. It is seen from Fig. 1.3 that the circuit accepts 4-bit inputs, and there is no additional information about the sign of the input number. If we assume that the port accepts non-negative integers, with 4-bits we can get integers in the range 0–15. Thus, we can write the **entity** segment as in PR 1.34.

PR 1.34 Program 1.34

```
entity FourBit_Circuit is
  port( inp1: in  integer range 0 to 15;
        inp2: in  integer range 0 to 15;
        outp1: out  integer range 0 to 31;
        outp2: out  integer range 0 to 31 );
end entity;
```

However, if we assume that the circuit port accepts both positive and negative integers, in this case, using 4-bits we can get the integers in the range -8 to 7, and using 5-bits we can get integers in the range -16 to 15.

Thus, we can write the **entity** segment as in PR 1.35,

PR 1.35 Program 1.35

```
entity FourBit_Circuit is
  port( inp1: in  integer range -8 to 7;
        inp2: in  integer range -8 to 7;
        outp1: out  integer range -16 to 15;
        outp2: out  integer range -16 to 15 );
end entity;
```

(d) The data type **natural** is used to indicate the non-negative integers. Using 4 bits we can represent the non-negative integers in the range 0–15. The **natural** is defined in the standard package.

Thus, we can write the **entity** segment as in PR 1.36.

PR 1.36 Program 1.36

```
entity FourBit_Circuit is
  port( inp1: in  natural range 0 to 15;
        inp2: in  natural range 0 to 15;
        outp1: out  natural range 0 to 31;
        outp2: out  natural range 0 to 31 );
end entity;
```

(e) The data type **positive** is used to indicate the positive integers. Using 4 bits we can represent the positive integers in the range 1–15. The **positive** is defined in the standard package.

Thus, we can write the **entity** segment as in PR 1.37.

PR 1.37 Program 1.37

```
entity FourBit_Circuit is
  port( inp1: in  positive range 1 to 15;
        inp2: in  positive range 1 to 15;
        outp1: out  positive range 1 to 31;
        outp2: out  positive range 1 to 31 );
end entity;
```

(f) The data type **unsigned** is used to represent the non-negative integers of any number of bits. The difference of this representation from that of the **natural** is that we decide on the number of bits to be used to represent the non-negative integers as a bit vector. To be able to use the **unsigned** data type in our VHDL programs we need to use **IEEE.numeric_std.all** package at our header.

Thus, we can write the **entity** segment as in PR 1.38.

PR 1.38 Program 1.38

```
library IEEE;
use IEEE.numeric_std.all;

entity FourBit_Circuit is
  port( inp1: in  unsigned(3 downto 0);
        inp2: in  unsigned (3 downto 0);
        outp1: out  unsigned (4 downto 0);
        outp2: out  unsigned (4 downto 0) );
end entity;
```

(g) The data type **signed** is used to represent the integers of any number of bits. The difference of this representation from that of the **integer** is that we decide on the number of bits to be used to represent integers as a bit vector. To be able to use the **signed** data type in our VHDL programs we need to use **IEEE. numeric_std.all** package at our header.

Thus, we can write the **entity** segment as in PR 1.39.

PR 1.39 Program 1.39

```
library IEEE;
use IEEE.numeric_std.all;

entity FourBit_Circuit is
  port( inp1: in  signed (3 downto 0);
        inp2: in  signed (3 downto 0);
        outp1: out  signed (4 downto 0);
        outp2: out  signed (4 downto 0) );
end entity;
```

Example 1.3 The black box representation of an electronic circuit is shown in Fig. 1.4. Describe the electronic circuit ports by a VHDL entity segment. Use **std_logic_vector** for the data types of the ports.

Fig. 1.4 The black box representation of an electronic circuit for Example 1.3

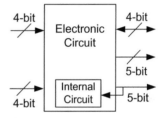

Solution 1.3 The black box illustration of the circuit shown in Fig. 1.4 has input, output, input/output and buffer ports. Considering the port types, we can write an **entitle** for the given circuit as in PR 1.40.

```
library IEEE;
use IEEE.std_logic_1164.all;

entity Electronic_Circuit is
  port( inp1: in std_logic_vector(3 downto 0);
        inp2: in std_logic_vector(3 downto 0);
        outp1: inout std_logic_vector(3 downto 0);
        outp2: out std_logic_vector(4 downto 0);
        outp3: buffer std_logic_vector(4 downto 0) );
end entity;
```

PR 1.40 Program 1.40

1.3 Architecture

The architecture part of a VHDL program describes the internal behavior of the electronic circuit. The data received from the ports of the electronic circuit is processed inside the architecture part and the output data is obtained. The produced output data is sent to the output ports of the electronic circuit. The general syntax of the architecture part is depicted in PR 1.41.

PR 1.41 Program 1.41

> **architecture** architecture_name **of** entity_name **is**
>   Declarations
> **begin**
>   Statements
> **end** [architecture] [architecture_name];

The last line of the **architecture** can be either "**end architecture**" or "**end** architecture_name" or "**end architecture** architecture_name" or just "**end**".

Example 1.4 The program segments given in PR 1.42 are equal to each other.

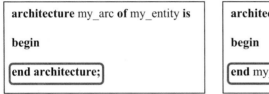

PR 1.42 Program 1.42

Example 1.5 The black box representation of a digital circuit is given in Fig. 1.5.

Fig. 1.5 The black box representation of a digital circuit for Example 1.5

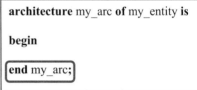

Write a VHDL program for the implementation of the circuit given in Fig. 1.5.

Solution 1.5 The entity part for the black box given in Fig. 1.5 can be written as in PR 1.43.

PR 1.43 Program 1.43

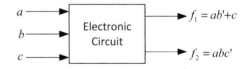

The behavior of the circuit can be expressed in architecture part. The architecture part of the circuit can be written considering the following steps.

(S1) First, write the reserved word architecture as in PR 1.44.

PR 1.44 Program 1.44

> **architecture**

(S2) In step 2, we give a name to architecture as in PR 1.45.

PR 1.45 Program 1.45

> **architecture** example_arc

(S3) In step 3, write the reserved word **of** to the end of the line of PR 1.45 as in PR 1.46.

PR 1.46 Program 1.46

> **architecture** example_arc **of**

(S4) In step 4, write name of the entity after the reserved word **of** as in PR 1.47.

PR 1.47 Program 1.47

> **architecture** example_arc **of** example_1_5

(S5) In step 5, write the reserved word **is** to the end of title name as in PR 1.48.

PR 1.48 Program 1.48

> **architecture** example_arc **of** example_1_5 **is**

(S6) In step 6, write the closing tag of architecture part as in PR 1.49.

PR 1.49 Program 1.49

> **architecture** example_arc **of** example_1_5 **is**
>
> **end** example_arc

(S7) In step 7, put a semicolon to the end of the closing tag as in PR 1.50.

PR 1.50 Program 1.50

> **architecture** example_arc **of** example_1_5 **is**
>
> **end** example_arc;

(S8) In step 8, write the reserved word **begin** as in PR 1.51.

PR 1.51 Program 1.51

> **architecture** example_arc **of** example_1_5 **is**
>
> **begin**
>
> **end** example_arc;

(S8) In step 8, we add the VHDL implementation of the Boolean functions $f_1 = ab' + c$ and $f_2 = abc'$ as in PR 1.52.

PR 1.52 Program 1.52

```
architecture example_arc of example_1_5 is

begin
    f1<= (a and not (b)) or c;
    f2<=  a and b and (not(c));
end example_arc;
```

With the entity part, our complete VHDL program becomes as in PR 1.53.

PR 1.53 Program 1.53

```
entity example_1_5 is
  port( a, b, c: in  bit;
            f1, f2: out  bit );
end entity;

architecture example_arc of example_1_5 is

begin
    f1<= (a and NOT (b)) or c;
    f2<=  a and b and (not(c));
end example_arc;
```

Note: In PR 1.53 the last line of the VHDL program can be written as "**end architecture**" or as "**end;**" For the name of the architecture part of a VHDL program, the names logic_flow, behavioral, behavior_of_the_circuit, circuit_logic, can be used.

Example 1.6 Write VHDL statements for the **Boolean** functions $f_1 = ab' + c$.

Solution 1.6 Let's implement the f_1 function as

$$f_1 <= a \text{ and } (\text{not}(b)) \text{ or } c.$$

However, this implementation is not complete, we have missing parenthesis around **and** logic. The correct statement is

$$f_1 <= (a \text{ and } (\text{not }(b))) \text{ or } c.$$

Example 1.7 A digital logic circuit is given in Fig. 1.6. Write a VHDL program for the implementation of the circuit given in Fig. 1.6.

Fig. 1.6 Digital logic circuit
or Example 1.7

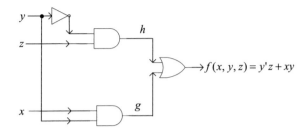

Solution 1.7 The title part for the given circuit can be written as in PR 1.54.

PR 1.54 Program 1.54

```
library IEEE;
use IEEE.std_logic_1164.all;

entity logic_circuit is
  port( x: in  std_logic;
        y: in  std_logic;
        z: in  std_logic;
        f: out  std_logic );
end logic_circuit;
```

The behavior of the circuit can be expressed in architecture part. The architecture part of the circuit can be written considering the following steps.

(S1) First, write the reserved word architecture as in PR 1.55.

PR 1.55 Program 1.55

```
architecture
```

(S2) In step 2, we give a name to architecture as in PR 1.56.

PR 1.56 Program 1.56

```
architecture logic_flow
```

(S3) In step 3, write the reserved word **of** to the end of the line of PR 1.56 as in PR 1.57.

PR 1.57 Program 1.57

```
architecture logic_flow of
```

(S4) In step 4, write name of the entity after the reserved word **of** as in PR 1.58.

PR 1.58 Program 1.58

```
architecture logic_flow of logic_circuit
```

(S5) In step 5, write the reserved word **is** to the end of title name as in PR 1.59.

PR 1.59 Program 1.59

```
architecture logic_flow of logic_circuit is
```

(S6) In step 6, write the closing tag of architecture part as in PR 1.60.

PR 1.60 Program 1.60

> **architecture** logic_flow **of** logic_circuit **is**
>
> **end** logic_flow

(S7) In step 7, put a semicolon to the end of the closing tag as in PR 1.61.

PR 1.61 Program 1.61

> **architecture** logic_flow **of** logic_circuit **is**
>
> **end** logic_flow;

(S8) In step 8, write the reserved word **begin** as in PR 1.62.

PR 1.62 Program 1.62

> **architecture** logic_flow **of** logic_circuit **is**
>
> **begin**
>
> **end** logic_flow;

(S8) In step 8, we add the VHDL implementation of the intermediate Boolean functions $g = xy$ and $h = y'z$ as in PR 1.63 Since, g and h are the internal outputs of the logic circuit, to hold the value of these outputs, we define two signal at the declarative part of the architecture as it is seen in PR 1.63. Note that the declarative part is the part between the words **architecture** and **begin**.

PR 1.63 Program 1.63

> **architecture** logic_flow **of** logic_circuit **is**
>   **signal** g, h: **std_logic**;
> **begin**
>   g<=x **and** y;
>   h<=**not**(y) **and** z;
> **end** logic_flow;

(S9) In step 9, we add the VHDL implementation of the Boolean function $f = g + h$ as in PR 1.64.

PR 1.64 Program 1.64

> **architecture** logic_flow **of** logic_circuit **is**
>   **signal** g, h: **std_logic**;
> **begin**
>   g<=x **and** y;
>   h<=**not**(y) **and** z;
>   f<=g **or** h;
> **end** logic_flow;

With the entity part, our complete VHDL program becomes as in PR 1.65.

```
library IEEE;
use
IEEE.std_logic_1164.all;

entity logic_circuit is
  port( x: in  std_logic;
         y: in  std_logic;
         z: in  std_logic;
         f: out std_logic );
end logic_circuit;
```

```
architecture logic_flow of logic_circuit is
  signal g, h: std_logic;
begin
  g<=x and y;
  h<=not(y) and z;
  f<=g or h;
end logic_flow;
```

PR 1.65 Program 1.65

1.4   Data Objects

In VHDL programming we have data objects which hold values. A data object has
a name and data type. The general syntax for the use of the data object is as

data object   object name :   data type := initial value;

where the data object can be **signal, variable,** or **constant**. The object name is up to
you. The data type can be any data type available in VHDL programming, such as
bit, **std_logic**, **integer**, **unsigned**, **signed**, **std_logic_vector**, **bit_vector**, etc.

The **signal** objects can be declared in the declarative part of the architecture. On
the other hand, the **variable** objects can be declared inside the sequential program
segments, like **process, procedure, function.**

The **constant** objects are usually declared in the declarative part of the **architecture, entity, package, process, procedure** and **function.** However, it can also
be declared in the package **body, block, generate.**

1.4.1   Constant Object

The value of the constant object cannot be changed through the program. Its syntax
is as follows

constant   name :   data type := initial value;

Example 1.8 Inspect the **constant** object declarations in PR 1.66.

---

constant my_constant: **integer**:=32;

constant my_flag: **std_logic**:='1';

constant my_vector : **std_logic_vector**(3 **downto** 0):= "1010";

PR 1.66 Program 1.66

Example 1.9 Declare a constant object which uses **unsigned** data type with initial value "100100", i.e., the length of the unsigned bit vector is 6.

Solution 1.9 We can declare the **constant** object in the question by the following steps.

(S1) First, write the reserved word **constant** as in PR 1.67.

PR 1.67 Program 1.67

---

constant

(S2) In step 2, we give a name to **constant** object as in PR 1.68.

PR 1.68 Program 1.68

---

constant my_number

(S3) In step 3, put a colon, i.e., :, to the end of the line as in PR 1.69.

PR 1.69 Program 1.69

---

constant my_number :

(S4) In step 4, write the type of the data to the end of the line as in PR 1.70.

PR 1.70 Program 1.70

---

constant my_number : **unsigned**(5 **downto** 0)

(S5) In step 5, put ':=' to the end of data type as in PR 1.71.

PR 1.71 Program 1.71

---

constant my_number: **unsigned**(5 **downto** 0):=

(S6) In step 6, add the initial value to the end of the line as in PR 1.72.

---

constant my_number: **unsigned**(5 **downto** 0):= "100100"

PR 1.72 Program 1.72

(S7) Finally, put a semicolon to the end of the line as in PR 1.73.

> **constant** my_number: **unsigned**(5 **downto** 0):= "100100";

PR 1.73 Program 1.73

Example 1.10 Declare a constant object which uses **natural** data type with initial value 2234566.

Solution 1.10 We can declare the **constant** object in the question by the following steps.

(S1) First, write the reserved word **constant** as in PR 1.74.

PR 1.74 Program 1.74

> **constant**

(S2) In step 2, we give a name to **constant** object as in PR 1.75.

PR 1.75 Program 1.75

> **constant** my_integer

(S3) In step 3, put a colon, i.e., :, to the end of the line as in PR 1.76.

PR 1.76 Program 1.76

> **constant** my_integer:

(S4) In step 4, write the type of the data to the end of the line as in PR 1.77.

PR 1.77 Program 1.77

> **constant** my_integer: **natural**

(S5) In step 5, put ':=' to the end of data type as in PR 1.78.

PR 1.78 Program 1.78

> **constant** my_integer: **natural** :=

(S6) In step 6, add the initial value to the end of the line as in PR 1.79.

PR 1.79 Program 1.79

> **constant** my_integer: **natural** := 2234566

(S7) Finally, put a semicolon to the end of the line as in PR 1.80.

PR 1.80 Program 1.80

> **constant** my_integer: **natural** := 2234566;

1.4.2 Signal Object

Signal object declarations may or may not include an initial value. Signal objects are usually employed in the declarative part of the **architectures,** and **packages**. Signal declarations are not allowed inside sequential program segments, such as **function, procedure**, and **process**. Its syntax is as follows

signal   name :   data type := initial value;

Example 1.11 Inspect the following **signal** object declarations. (PR 1.81)

> **signal** my_number: **integer**;
>
> **signal** my_bit: **bit**:='1';
>
> **signal** my_vector : **std_logic_vector**(3 **downto** 0):= "1010";

PR 1.81 Program 1.81

Example 1.12 Declare a **signal** object which uses **signed** data type for 6-bit integers.

Solution 1.12 We can declare the **signal** object using the following steps.
   (S1) First, write the reserved word **signal** as in PR 1.82.

PR 1.82 Program 1.82

> **signal**

   (S2) In step 2, we give a name to **signal** object as in PR 1.83.

PR 1.83 Program 1.83

> **signal** my_integer

   (S3) In step 3, put a colon, i.e., :, to the end of the line as in PR 1.84.

PR 1.84 Program 1.84

> **signal** my_integer:

   (S4) In step 4, write the type of the data to the end of the line as in PR 1.85.

PR 1.85 Program 1.85

> **signal** my_integer: **signed**(5 **downto** 0)

(S7) Since no initial value is indicated in the question, we can terminate the declaration by putting a semicolon to the end of the line as in PR 1.86.

PR 1.86   Program 1.86

> **signal** my_integer: **signed** (5 **downto** 0);

1.4.3   Variable Object

Variable object declarations may or may not include an initial value. Variable objects are employed in sequential program units, such as **function, procedure**, and **process**. Its syntax is as follows

> **variable**   name :   data type := initial value;

Example 1.13 Inspect the **variable** object declarations in PR 1.87.

> **variable** my_number: **natural**;
>
> **variable** my_logic: **std_logic**:='1';
>
> **variable** my_vector : **std_logic_vector**(3 downto 0):= "1110";

PR 1.87   Program 1.87

Example 1.14 Declare a **variable** object which uses **integer** data type with initial value 124542.

Solution 1.14 We can declare the **variable** object in the question by the following steps.
(S1) First, write the reserved word **variable** as in PR 1.88.

PR 1.88   Program 1.88

> **variable**

(S2) In step 2, we give a name to **variable** object as in PR 1.89.

PR 1.89   Program 1.89

> **variable** my_integer

(S3) In step 3, put a colon, i.e., :, to the end of the line as in PR 1.90.

PR 1.90   Program 1.90

> **variable** my_integer:

(S4) In step 4, write the type of the data to the end of the line as in PR 1.91.

PR 1.91 Program 1.91

> **variable** my_integer: **integer**

(S5) In step 5, assign the initial value of the data type putting ': = 124542' to the end of data type as in PR 1.92.

PR 1.92 Program 1.92

> **variable** my_integer: **integer** :=124542

(S6) In step 6, we terminate the declaration by putting a semicolon to the end of the line as in PR 1.93.

PR 1.93 Program 1.93

> **variable** my_integer: **integer** :=124542;

Example 1.15 The architecture of a VHDL program is given as in PR 1.94. Indicate the declarative part of the given architecture segment.

PR 1.94 Program 1.94

> **architecture** logic_flow **of** logic_circuit **is**
>   **signal** g, h: **std_logic**;
>   **signal** k: **integer;**
> **begin**
>   g<=x **and** y;
>   h<=**y xor** z;
>   k<=1000;
> **end** logic_flow;

Solution 1.15 The declarative part of the architecture is the part between the reserved word **architecture** and **begin**. Considering this information, we can indicate the declarative part of the **architecture** as in PR 1.95.

PR 1.95 Program 1.95

> **architecture** logic_flow **of** logic_circuit **is**
>   **signal** g, h: **std_logic**;   Declarative
>   **signal** k: **integer;**   Part
> **begin**
>   g<=x **and** y;
>   h<=**y xor** z;
>   k<=1000;
> **end** logic_flow;

1.5   VHDL Operators

The operators provided by VHDL can be classified in the following main categories.

- Assignment Operators
- Logical Operators
- Relational Operators
- Arithmetic Operators
- Concatenation Operator

Let's briefly explain all those operators.

1.5.1   Assignment Operators

Operator ":="

We use operator ":=" for either initial value assignment or to assign a value to a variable, constant or generic.

Note that ":=" is used for signal objects only for initial value assignment. It is not used for signal objects in any other cases.

Example 1.16 **signal** my_signal: **integer:=**10; – Initial value assignment
my_signal <=30; – Value assignment

Example 1.17 **variable** my_signal: **integer:=**10; – Initial value assignment
my_signal:=40; – Value assignment

Example 1.18 **constant** my_number: **natural:=**214; – Initial value assignment

Operator "<="

We use the operator "<=" to assign a value to a signal object.

Example 1.19 **signal** my_signal: **bit_vector(3 downto 0):=** "10011"; – Initial value assignment
my_signal<= "11011"; – Value assignment

Operator "=>" and "**others**" keyword

We use operator "=>" to assign values to vector elements. The operator "=>" is usually employed with the reserved word **others** which indicates the index values of the unassigned vector elements.

Example 1.20 After the program line
signal x_vec: **std_logic_vector**(5 downto 1): =(4|3=> '1', others=> '0');
we have x_vec="01100", and after the line
x_vec<=(1 to 3=>'1', others=>'0');
we have x_vec="00111",

Example 1.21 After the program line
   variable   x_vec: **std_logic_vector**(0   to   10): =(1|3|5=>   '1',   2=>   'Z',
others=>'0');
   we have x_vec= "01Z10100000", and after the line
   x_vec:=(1 to 5=>'0', others=>'Z');
   we have x_vec="Z00000ZZZZZ",

1.5.2   Logical Operators and Shift Operators

The logical operators are used for data types **bit**, **std_logic**, **std_logic_vector**,
std_ulogic, **std_ulogic_vector**, and **Boolean**. Logical operators are used for the
implementation combinational logic circuits. The logical operators and shift operators can be summarized as in Table 1.3.

Table 1.3 Logical and shift
operators

Logical operators	**and**, **or**, **nand**, **nor**, **xor**, **xnor**, **not**
Shift operators	**sll**, **srl**, **sla**, **sra**, **rol**, **ror**

   Let's briefly explain the logical shift operators briefly.
   Shift Operators
   Let's briefly explain the logical shift operators briefly.
   SLL: Shift Left Logical
   The bits are shifted to the left and new positions are filled by zeros.

Example 1.22 If

$$x < = '10101111'$$

then

$$y < = x \, \textbf{sll} \, 2$$

produces

$$y < = '10111100'.$$

   SLA: Shift Left Arithmetic
   The bits are shifted to the left and new positions are filled by the value of the
rightmost bit.

Example 1.23 If

$$x < = '11111110'$$

then

$$y <= x \, \textbf{sla} \, 3$$

produces

$$y <= {}'11110000'.$$

SRL: Shift Right Logical
The bits are shifted to the right and new positions are filled by zeros.

Example 1.24 If

$$x <= {}'10101111'$$

then

$$y <= x \, \textbf{srl} \, 2$$

produces

$$y <= {}'00101011'.$$

SRA: Shift Right Arithmetic
The bits are shifted to the right and new positions are filled by the value of the leftmost bit.

Example 1.25 If

$$x <= {}'01111111'$$

then

$$y <= x \, \textbf{sra} \, 3$$

produces

$$y <= {}'00001111'.$$

ROL: Rotate Left
The bits are shifted towards left, and the new positions on the right are filled with the bits dropped from the left edge.

Example 1.26 If

$$x <= {}'01111111'$$

then

$$y <\, = x\,\textbf{rol}\,3$$

produces

$$y <\, = '11111011'.$$

ROR: Rotate Right

The bits are shifted towards right, and the new positions on the left are filled with the bits dropped from the right edge.

Example 1.27 If

$$x <\, = '111111000'$$

then

$$y <\, = x\,\textbf{ror}\,3$$

produces

$$y <\, = '000111111'.$$

Note: Multiplying an unsigned integer by two is the same as shifting the number to the left by one bit. In a similar manner, dividing an unsigned integer by two is the same as shifting the number to the right by one bit.

Example 1.28 Write a VHDL program described as follows. The program inputs an 8-bit number. Let's denote this number by x. The program sends the numbers are $y_1 = 2^4 \times x$, $y_2 = 2^{-3} \times x$ to its ports.

Solution 1.28 With the given information, we can write the entity part of the program as in PR 1.96.

PR 1.96 Program 1.96

```
entity power_two_operations is
  port( inp: in  bit_vector(7 downto 0);
              outp1: bit_vector(7 downto 0);
              outp2: bit_vector(7 downto 0) );
end power_two_operations;
```

The operations $y_1 = 2^4 \times x$, $y_2 = 2^{-3} \times x$ can be achieved using the shift operators **shl** and **shr** as in PR 1.97.

PR 1.97 Program 1.97

```
architecture logic_flow of power_two_operations is
begin
  outp1<=x sll 4;
  outp2<=x srl 3;
end logic_flow;
```

1.5.3 Relational Operators

Relational operators are used for comparison purposes. The following relational operators can be used in VHDL programming are

$$= \quad \rightarrow \quad \textit{Equal to}$$
$$\neq \quad \rightarrow \quad \textit{Not equal to}$$
$$> \quad \rightarrow \quad \textit{Greater than}$$
$$< \quad \rightarrow \quad \textit{Less than}$$
$$>= \quad \rightarrow \quad \textit{Greater than or equal to}$$
$$<= \quad \rightarrow \quad \textit{Less than or equal to.}$$

The result of the relational operator is a Boolean value, i.e., true or false.

1.5.4 Arithmetic Operators

Arithmetic operators are used to do arithmetic operations on integers or numeric values. The arithmetic operators that can be used in VHDL programming are

$$+ \quad \rightarrow \quad \textit{Addition}$$
$$- \quad \rightarrow \quad \textit{Subtraction}$$
$$* \quad \rightarrow \quad \textit{Multiplication}$$
$$/ \quad \rightarrow \quad \textit{Division}$$
$$** \quad \rightarrow \quad \textit{Exponentiation}$$
$$\textbf{mod} \quad \rightarrow \quad \textit{Modulus}$$
$$\textbf{rem} \quad \rightarrow \quad \textit{Remainder}$$
$$\textbf{abs} \quad \rightarrow \quad \textit{Absolutevalue.}$$

Let's explain these operators shortly:
Operator $**$
The operation

$$y = x^a$$

can be implemented in VHDL using

$$y <= x ** a \quad \text{or} \quad y := x ** a.$$

Exponentiation operator is synthesizable for $x = 2$ and integer a values, or it is synthesizable when $a = 2$. For other cases, it is not synthesizable.

Example 1.29 Write a VHDL program that calculates $y = 2^x$ where x is an integer in the range 0–7.

Solution 1.29 The function $y = 2^x$ can be implemented in VHDL as shown in PR 1.98.

PR 1.98 Program 1.98

```
entity power_of_2_circuit is
  port( x: in  integer range 0 to 7;
           y: out  integer range 1 to 128);
end entity;

architecture logic_flow of power_of_2_circuit is

begin
    y<=2**x;
end architecture;
```

Operator mod

The result of $x \bmod y$ is the remainder of x/y, and the remainder has the sign of y. The operator **mod** is synthesizable without any restrictions.

Example 1.30 For the **mod** operator we can give the following examples

$$10 \bmod 3 = 1 \qquad 10 \bmod -3 = -2$$
$$-10 \bmod 3 = 2 \qquad -10 \bmod -3 = -1.$$

Example 1.31 Write a VHDL program that takes a four-bit integer from the port, and calculates the remainder after division by 3, and displays the remainder at the output port.

Solution 1.31 The function $y = x \bmod 3$ can be implemented in VHDL as shown in PR 1.99.

PR 1.99 Program 1.99

```
entity mod_by_3_circuit is
  port( x: in  integer range -8 to 7;
           y: out integer range 0 to 2);
end entity;

architecture logic_flow of mod_by_3_circuit is

begin
    y<=x mod 3;
end architecture;
```

Operator rem

The result of $x\,rem\,y$ is the remainder of x/y, and the remainder has the sign of x. The operator **rem** is synthesizable.

Example 1.32

$$10\,rem\,3 = 1 \qquad 10\,rem\,-3 = 1$$
$$-10\,rem\,3 = -1 \qquad -10\,rem\,-3 = -1$$

Example 1.33 Write a VHDL program that takes a four-bit integer from the port, and calculates the remainder after division by 3, and displays the remainder at the output port.

Solution 1.33 The function $y = x\,rem\,3$ can be implemented in VHDL as shown in PR 1.100.

PR 1.100 Program 1.100

```
entity rem_by_3_circuit is
  port( x: in integer range -8 to 7;
        y: out integer range 0 to 2);
end entity;

architecture logic_flow of rem_by_3_circuit is

begin
  y<=x rem 3;
end architecture;
```

Operator abs

The **abs** operator checks the most significant bit of the number, and if it is '1', then two's complement of the number is evaluated, otherwise, the number is kept as it is.

The **abs** operator has been implemented for different arguments in different packages. These packages and their **abs** implementations are summarized as follows:

Package: **std_logic_arith**, have the functions:
function "abs"(L: **signed**) **return signed**;
function "abs"(L: **signed**) **return std_logic_vector**;
Package: **numeric_bit**, have the function.
function "abs" (arg: **signed**) **return signed**;
Package: **std_logic_signed**, have the function:
function "abs"(L: **std_logic_vector**) **return std_logic_vector**;
Package: **numeric_std**, have the function:
function "abs" (arg: **signed**) **return signed**;

Example 1.34 Write a VHDL program that takes two eight-bit signed integers x and y from the port, and calculates $z = |x| + |y|$ and sends the result to an output port.

Solution 1.34 The function $z = |x| + |y|$ can be implemented in VHDL as shown in PR 1.101.

```
library ieee;
use ieee.numeric_std.all;
-- Note that "numeric_std" package includes "std_logic_1164" package in its
implementation

entity abs_sum_circuit is
  port( x: in  signed(7 downto 0);
        y: in  signed(7 downto 0);
        z: out signed(8 downto 0) );
end entity;

architecture logic_flow of abs_sum_circuit is

begin
  z<=abs(x)+abs(y);
end architecture;
```

PR 1.101   Program 1.101

Example 1.35 If I edit the package file **std_logic_signed.vhd**, I see the following line inside the file code:

$$\text{function ``abs''}(L : \textbf{std\_logic\_vector}) \textbf{ returnstd\_logic\_vector}$$

What does this line mean?

Solution 1.35 This line means that you can use **abs** operator in your VHDL program as in the form

$$\text{std\_logic\_vector} < = \text{abs}(\text{std\_logic\_vector})$$

if you include **std_logic_signed** package as header file in your VHDL program.

1.5.5   Concatenation Operator "&"

Concatenation operator is used to merge logic or bit vectors to get longer logic vector. The concatenation operator can be used to merge the data types, **bit_vector, std_logic_vector, std_ulogic_vector, signed, unsigned, string, integer vector, boolean-vector.**

Example 1.36 Write a VHDL program that takes two bit-vectors of length 5 and 7, concatenates them and sends the result to an output port.

Solution 1.36 We can write the required VHDL program as in PR 1.102.

PR 1.102 Program 1.102

```
entity merger_two_vectors is
  port( x_vec: in  bit_vector(4 downto 0);
            y_vec: in  bit_vector(6 downto 0);
            z_vec: out bit_vector(11 downto 0) );
end entity;

architecture logic_flow of merger_two_vectors is

begin
    z_vec<=x_vec & y_vec;
end architecture;
```

Example 1.37 Another example for the concatenation operator is given in PR 1.103.

PR 1.103 Program 1.103

```
constant x: std_logic:='Z';

signal y: std_logic_vector(4 downto 0);

y<=x&'1'&x&"01";
```

1.5.6 Generic Statement

The keyword **generic** is used to declare generic parameters which can be used through the VHDL program. In this way, it becomes easier to adapt the VHDL program to different applications. The **generic** declarations are done in the **entity** part before the **port** section. The syntax for **generic** declarations is as follows:

$$\textbf{generic}(\text{parameter name} - 1 : \text{data type} - 1 := \text{initial value} - 1;$$
$$\text{parameter name} - 2 : \text{data type} - 2 := \text{initial value} - 2;$$
$$\cdots$$
$$\text{parameter name} - N : \text{data type} - N := \text{initial value} - N);$$

Example 1.38 An example use of **generic** declarations is shown in PR 1.104. Note that **generic** declarations are made before the **port** declarations.

PR 1.104 Program 1.104

```
entity example_generic is

    generic (N : natural:=8;
             M: natural:=16 );

    port( x_vec: in  bit_vector(N-1 downto 0);
          y_vec: in  bit_vector(M-1 downto 0);
          z_vec: out bit );

end entity;
```

Example 1.39 Declare any two logic vectors as generic parameters.

Solution 1.39 We can declare two generic logic vectors as follows:
(S1) First, write the reserved word entity as in PR 1.105

PR 1.105 Program 1.105

```
generic
```

(S2) In step 2, add parentheses to the **generic** as in PR 1.106.

PR 1.106 Program 1.106

```
generic()
```

(S3) In step 3, put a semicolon to the end of the line as in PR 1.107.

PR 1.107 Program 1.107

```
generic();
```

(S4) In step 4, write the name of the generic parameters as in PR 1.108.

PR 1.108 Program 1.108

```
generic(vec_1
        vec2);
```

(S5) In step 5, put colons after parameter names and write the data type as in PR 1.109.

PR 1.109 Program 1.109

```
generic(vec_1: std_logic_vector(7 downto 0)
        vec_2: std_logic_vector(3 downto 0) );
```

(S6) In step 6, assign values to the generic vectors as in PR 1.110.

```
generic(vec_1: std_logic_vector(7 downto 0):="00000000"
        vec_2: std_logic_vector(3 downto 0):="1010" );
```

PR 1.110 Program 1.110

(S7) In step 7, put a semicolon after the first generic declaration as in PR 1.111. Note that after the last generic declaration, we do not use a semicolon at the end of the line.

```
generic(vec_1: std_logic_vector(7 downto 0):="00000000";
        vec_2: std_logic_vector(3 downto 0):="1010" );
```

PR 1.111 Program 1.111

(S8) Lastly, don't forget that we place the generic declarations before the **port** section of the **entity** part, as in PR 1.112.

```
library ieee;
use ieee. std_logic_1164.all;

entity example_generic is

generic(vec_1: std_logic_vector(7 downto 0):="00000000";
        vec_2: std_logic_vector(3 downto 0):="1010" );

  port( inp: in std_logic;
        outp: out std_logic);
end entity;
```

PR 1.112 Program 1.112

Problems

(1) The black box representation of an electronic circuit is shown in Fig. 1.P1. Describe the electronic circuit ports by a VHDL program.

Fig. 1.P1 Black box representation of an electronic circuit

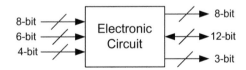

(a) Use the data type **std_logic_vector** to describe the ports in **entity** part.
(b) Use the data type **bit_vector** to describe the ports in **entity** part.
(c) Use the data type **integer** to describe the ports in **entity** part.
(d) Use the data type **natural** to describe the ports in **entity** part.
(e) Use the data type **positive** to describe the ports in **entity** part.

(f) Use the data type **unsigned** to describe the ports in **entity** part.
(g) Use the data type **signed** to describe the ports in **entity** part.

(2) Implement the **Boolean** function $f = ab' + bc'$ in VHDL.
(3) Define a **signal** object with data type **integer**, and the initial value of the object is 2342.
(4) Consider the **signal** object "**signal** my_obj: **integer;**". How many bits are used for this signal object?
(5) State the differences between a signal object and variable object.
(6) Find the results of the following operations

$$-15\,mod\,3 \quad 15\,mod\,-3 \quad -15\,mod\,-3$$
$$-15\,rem\,3 \quad 15\,rem\,-3 \quad -15\,rem\,-3.$$

(7) Declare two generic signal objects having integer data types with initial values 8 and 15.

Chapter 2
Combinational Logic Circuit Design and Concurrent Coding in VHDL

In this chapter we will explain the implementations of combinational logic circuits in VHDL language. The combinational logic circuits can be implemented using logic operators and VHDL statements. The VHDL statements **when** and **select** are used for the implementation if conditional expressions are available in logic circuit implementation. The combinational circuits are implemented via concurrent codes.

2.1 "When" and "Select" Statements

The VHDL statements **when** and **select** can be employed in concurrent VHDL codes. These statements are not used in sequential, i.e., clocked, program units. We can employ VHDL statements **when** and **select** for the implementation of combinational circuits. These statements are employed for the conditional implementation of the logic circuits. The syntax of the **when** statement is as follows:

$$< \text{signal object} > \quad <= \quad < \text{statement} > \text{ when } < \text{condition} > \text{ else}$$
$$< \text{statement} > \text{ when } < \text{condition} > \text{ else}$$
$$\ldots$$
$$< \text{statement} > \text{ when } < \text{condition} > \text{ else}$$
$$< \text{statement} > ;$$

The **when** and **select** statements are similar to each other. They differ only in implementation syntax. The syntax of the **select** statement is as follows:

$$\textbf{with} < \text{condition} > \textbf{select}$$
$$< \text{signal object} > \quad <= \quad < \text{statement} > \textbf{ when} < \text{condition} > ,$$
$$< \text{statement} > \textbf{ when} < \text{condition} > ,$$
$$\ldots$$
$$< \text{statement} > \textbf{ when} \text{others};$$

© Springer Nature Singapore Pte Ltd. 2019
O. Gazi, *A Tutorial Introduction to VHDL Programming*,
https://doi.org/10.1007/978-981-13-2309-6_2

Let's now give some examples illustrating the combinational synthesizable logic circuit design using VHDL programming.

Example 2.1 Implement the Boolean function $f(x, y, z) = x'y' + y'z$ using logical operators.

Solution 2.1 The Boolean function $f(x, y, z) = x'y' + y'z$ can be implemented using logical operators as

$$f <= (\textbf{not}(x)\textbf{and not}(y))\textbf{or}(\textbf{not}(y)\textbf{and } z)$$

A VHDL program that implements the Boolean function $f(x, y, z) = x'y' + y'z$ can be written as in PR 2.1.

```
entity f_function is
  port( x, y, z: in  bit;
            f: out bit );
  end entity;

architecture logic_flow of f_function is

begin
      f<=(not(x) and not(y)) or (not(y) and z);
  end architecture;
```

PR 2.1 Program 2.1

Example 2.2 The truth table of a Boolean function is given in Table 2.1.

Table 2.1 Truth table of a Boolean function

x	y	z	$f(x, y, z)$
0	0	0	1
0	0	1	1
0	1	0	0
0	1	1	0
1	0	0	0
1	0	1	1
1	1	0	0
1	1	1	0

Implement the Boolean function using **when** and **select** statements.

Solution 2.2 With **when** statement, it can be implemented as

$$f <= '1' \textbf{ when}(x =' 0' \textbf{ and } y =' 0' \textbf{ and } z =' 0')\textbf{else}$$
$$'1' \textbf{ when}(x =' 0' \textbf{ and } y =' 0' \textbf{ and } z =' 1')\textbf{else}$$
$$'1' \textbf{ when}(x =' 1' \textbf{and } y =' 0' \textbf{ and } z =' 1')\textbf{else}$$
$$'0';$$

With **select** statement, it can be implemented as

$$\textbf{with}(x\&y\&z) \textbf{ select}$$
$$f <='1' \textbf{ when } "000",$$
$$'1' \textbf{ when } "001",$$
$$'1'\textbf{when } "101",$$
$$'0' \textbf{ when others};$$

Example 2.3 Implement the Boolean function $f(x, y, z) = x'y' + y'z$ using **when** statement.

Solution 2.3 First, let's construct the truth table of the Boolean function $f(x, y, z) = x'y' + y'z$. If the Boolean function $f(x, y, z) = x'y' + y'z$ is inspected in details, we see that the function takes the value of 1 for $x = y = 0$ or for $y = 0, z = 1$. Considering this information, we can make the truth table of the given Boolean function as in Table 2.2.

Table 2.2 Truth table of $f = x'y' + y'z$

x	y	z	$f(x, y, z)$
0	0	0	1
0	0	1	1
0	1	0	0
0	1	1	0
1	0	0	0
1	0	1	1
1	1	0	0
1	1	1	0

Regarding the truth table, we can write VHDL program with **when** statement as in PR 2.2.

```
entity f_function is
  port( x, y, z: in  bit;
        f: out bit );
end entity;

architecture logic_flow of f_function is

begin
    f<='1' when (x='0' and y='0' and z='0') else
        '1' when (x='0' and y='0' and z='1') else
        '1' when (x='1' and y='0' and z='1') else
        '0';
end architecture;
```

PR 2.2 Program 2.2

The program in PR 2.2 can also be written as in PR 2.3.

```
entity f_function is
  port( xyz: in  bit_vector(2 downto 0);
        f: out bit );
end entity;

architecture logic_flow of f_function is

begin
    f<='1' when (xyz(2)='0' and xyz(1)='0' and xyz(0)='0') else
        '1' when (xyz(2)='0' and xyz(1)='0' and xyz(0)='1') else
        '1' when (xyz(2)='1' and xyz(1)='0' and xyz(0)='1') else
        '0';
end architecture;
```

PR 2.3 Program 2.3

Example 2.4 We can implement the Boolean function $f(x, y, z) = x'y' + y'z$ using **select** statement as in PR 2.4.

PR 2.4 Program 2.4

```
entity f_function is
  port( x, y, z: in  bit;
             f: out bit );
end entity;

architecture logic_flow of f_function is

begin
  with (x&y&z) select
        f<='1' when "000",
           '1' when "001",
           '1' when "101",
           '0' when others;
end architecture;
```

Example 2.5 Implement the function in (2.1) using VHDL.

$$f(x) = \begin{cases} 2 & \text{if } x = 1 \text{ or } x = 2 \\ 4 & \text{if } 3 \le x \le 6 \\ 0 & \text{otherwise.} \end{cases} \qquad (2.1)$$

Solution 2.5 The real valued function given in the question can be implemented in VHDL as in PR 2.5.

PR 2.5 Program 2.5

```
entity fx_function is
  port( x: in  integer;
        y: out integer range 0 to 4);
end entity;

architecture logic_flow of fx_function is

begin
  with x select
    y <= 2 when 1 | 2,
         4 when 3 to 6,
         0 when others;
end architecture;
```

2.2   Generate Statement

The VHDL statement **generate** is a concurrent statement. It is used to generate multiple instances of a program segment. The **generate** statement has two different forms which are unconditional **generate** and conditional **generate.** Let's see these forms separately.

Unconditional Generate

The unconditional **generate** as its name implies has no conditional part. Its syntax is as follows:

Label : **for** parameter **in** number − range **generate**
[declarative part
begin]
Statements
end generate[Label];

The use of Label is a must in **generate** statement, and the word **begin** is used if declarative part is available in the generate statement.

Example 2.6 The generate statement in

Label 1 : **for** indx **in** 0 to 3 **generate**
$y(\text{indx}) <= x(\text{indx})\textbf{xor}\, x(\text{indx} + 1)$;
end generate;

is the same as the program segment

$y(0) <= x(0)\textbf{xor}\, x(1)$;
$y(1) <= x(1)\textbf{xor}\, x(2)$;
$y(2) <= x(2)\textbf{xor}\, x(3)$;
$y(3) <= x(3)\textbf{xor}\, x(4)$;

That is, the same digital logic circuit synthesized for both program segments.

Example 2.7 Write a VHDL statement that reverses the elements of an 8-bit logic vector.

Solution 2.7 Let x and y be an 8-bit vectors. We can reverse the content of vector x using the following VHDL statement:

reverse : **for** indx **in** 0 to 7 **generate**
$y(\text{indx}) <= x(7 - \text{indx})$;
end generate;

Example 2.8 Write the previous example as a complete VHDL program.

Solution 2.8 The complete VHDL program is given in PR 2.6.

PR 2.6 Program 2.6

```
entity reverse_vector is
  port( x_vec: in  bit_vector(7 downto 0);
        y_rvec: out bit_vector(7 downto 0));
end entity;

architecture logic_flow of reverse_vector is

begin

  reverse: for indx in 0 to 7 generate
          y_rvec(indx)<=x_vec(7-indx);
  end generate;

end architecture;
```

Parity Generator

The parity of the binary vector

$$\bar{x} = [x_0 x_1 x_2 \cdots x_{N-1}] \qquad (2.2)$$

is calculated as

$$p = x_0 \oplus x_1 \oplus \cdots \oplus x_{N-1} \qquad (2.3)$$

where $\oplus$ is the **xor** operation. Note that, **xor** is an odd function, i.e., if the number of '1's in binary vector $\bar{x}$ is an odd integer, then $p = 1$, otherwise $p = 0$.

Example 2.9 Write a VHDL statement that calculates the parity for 16-bit logic vector.

Solution 2.9 Let x_vec be a 16-bit vector, and parr_vec be another 16-bit vector. We can calculate the parity bit for x_vec using the following VHDL statement:

```
parr_vec(0) <=x(0);

parity : for indx in 1 to 15 generate
        parr_vec(indx) <=parr_vec(indx − 1)xor x_vec(indx);
end generate;

p<=parr_vec(15);
```

Example 2.10 Write the previous example as a complete VHDL program.

Solution 2.10 The complete VHDL program is given in PR 2.7.

```
entity parity_bit_generator is
  port( x_vec: in  bit_vector(15 downto 0);
         p_bit: out bit);
end entity;

architecture logic_flow of parity_bit_generator is

  signal parr_vec: bit_vector(15 downto 0);

begin
  parr_vec(0)<= x_vec(0);

  parity: for indx in 1 to 15 generate
           parr_vec(indx)<= parr_vec(indx-1) xor x_vec(indx);
  end generate;
  p_bit<= parr_vec(15);
end architecture;
```

PR 2.7 Program 2.7

Conditional Generate

The syntax of the conditional generate is as

> Label : **if** condition **generate**
>   [declarative_part
>   **begin**]
>   VHDL Statements
>   **end generate**[Label];

where 'condition' must be a static expression.

   Note: Note that in VHDL syntax [···] shows the optional part.

Example 2.11 Write a VHDL statement that checks a constant positive integer and detects whether it is an even or odd integer.

Solution 2.11 The VHDL program segment can be written as follows:

> EvenDetector : **if**((number **mod** 2) = 0)**generate**
>   EvenFlag < =1;
>   **end generate**;
> OddDetector : **if**((number **mod** 2) = 1)**generate**
>   EvenFlag < =0;
>   **end generate**;

Example 2.12 Write the previous example as a complete VHDL program.

Solution 2.12 The complete program is depicted in PR 2.8.

```
entity even_odd_detector is
  port( EvenFlag: out bit);
end entity;

architecture logic_flow of even_odd_detector is
  constant number: positive:=10001; -- 31-bit positive integer
  begin
      EvenDetector: if ((number mod 2)=0) generate
          EvenFlag<=1;
      end generate;
      OddDetector: if ((number mod 2)=1) generate
          EvenFlag<=0;
      end generate;
end architecture;
```

PR 2.8 Program 2.8

Let's see the VHDL implementation of some well-known combinational circuits which do not require a clock for the operation of the circuits.

2.3 Examples for Combinational Circuits Implemented in VHDL

In this section we will see the implementation of some well-known combinational circuits in VHDL.

Multiplexer

The black box representation of a 2 × 1 multiplexer is depicted in Fig. 2.1.

Fig. 2.1 2 × 1 Multiplexer

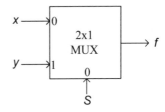

The multiplexer can be considered as a traffic regulator for digital circuits. The truth table of a 2 × 1 multiplexer is given in Table 2.3.

Table 2.3 Truth Table of the 2 × 1 Multiplexer

S	f		x	y	S	f
0	x	$\rightarrow$	0	0	0	0
1	y		0	0	1	0
			0	1	0	0
			0	1	1	1
			1	0	0	1
			1	0	1	0
			1	1	0	1
			1	1	1	1

Using the truth table given in Table 2.3, we can express the output of the 2 × 1 multiplexer circuit in terms of the selection and input logics as

$$f = S'x + Sy. \tag{2.4}$$

The internal circuit diagram of the black box in Fig. 2.1 is shown in Fig. 2.2.

Fig. 2.2 2 × 1 Multiplexer circuit diagram

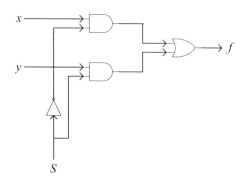

There are a number of ways to implement a multiplexer in VHDL. Let's see the implementation of a 2 × 1 multiplexer in VHDL.

Example 2.13 Implement 2 × 1 multiplexer in VHDL.

Solution 2.13 Considering Table 2.1, the output of the multiplexer can be expressed in terms of selection and input logics as

$$f = S'x + Sy. \tag{2.5}$$

Using the expression in (2.5), we can implement the 2×1 multiplexer as in PR 2.9.

PR 2.9 Program 2.9

```
library ieee;
use ieee.std_logic_1164.all;

entity multiplexer_2x1 is
  port( x, y, s: in std_logic;
           f: out std_logic );
end entity;

architecture logic_flow of multiplexer_2x1 is

begin
        f<=(not s and x) or (s and y);
end architecture;
```

Using the **when** VHDL statement we can implement 2×1 multiplexer as in PR 2.10.

PR 2.10 Program 2.10

```
library ieee;
use ieee.std_logic_1164.all;

entity multiplexer_2x1 is
  port( x, y, s: in std_logic;
           f: out std_logic );
end entity;

architecture logic_flow of multiplexer_2x1 is

begin
        f<=x when s='0' else
           y ;
end architecture;
```

Or using the explicit form of the truth table in Table 2.3, we can implement the 2×1 multiplexer using **when** statement as in PR 2.11.

PR 2.11 Program 2.11

```
library ieee;
use ieee.std_logic_1164.all;

entity multiplexer_2x1 is
  port( x, y, s: in std_logic;
        f: out std_logic );
end entity;

architecture logic_flow of multiplexer_2x1 is

begin
   f<='1' when (x='0' and y='1' and s='1') else
      '1' when (x='1' and y='0' and s='0') else
      '1' when (x='1' and y='1' and s='0') else
      '1' when (x='1' and y='1' and s='1') else
      '0';
end architecture;
```

If we use logic vector at the input port, we can write PR 2.11 as in PR 2.12.

PR 2.12 Program 2.12

```
library ieee;
use ieee.std_logic_1164.all;

entity multiplexer_2x1 is
  port( xys: in std_logic_vector(2 downto 0);
        f: out std_logic );
end entity;

architecture logic_flow of multiplexer_2x1 is

begin
   f<='1' when (xys="011") else
      '1' when (xys="100") else
      '1' when (xys="110") else
      '1' when (xys="111") else
      '0';
end architecture;
```

Using the **select** statement, we can implement the 2×1 multiplexer as in PR 2.13.

PR 2.13   Program 2.13

```
library ieee;
use ieee.std_logic_1164.all;

entity multiplexer_2x1 is
  port( x, y, s: in std_logic;
            f: out std_logic );
end entity;

architecture logic_flow of multiplexer_2x1 is

begin
  with (s) select
        f<=x when '0',
           y when others;
end architecture;
```

Example 2.14 Implement the digital circuit given in Fig. 2.3 in VHDL.

Fig. 2.3 Digital circuit for Example-2.14

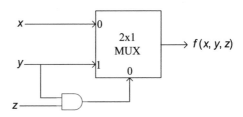

Solution 2.14 We can write a Boolean function for the output of the multiplexer in terms of the circuit inputs, and use the written function for the implementation of the circuit in VHDL language. At the selection line of the multiplexer, we have the Boolean function $S = yz$. The operation of the multiplexer can be described as

$$S = 0 \rightarrow f = x$$
$$S = 1 \rightarrow f = y. \qquad (2.6)$$

Then, the output of the multiplexer can be written as $f = S'x + Sy$ in which using $S = yz$ we get

$$f = (yz)'x + yzy \rightarrow f = xy' + xz' + yz. \qquad (2.7)$$

And using the obtained Boolean function, we can implement the circuit given in Fig. 2.3 as in PR 2.14.

```
library ieee;
use ieee.std_logic_1164.all;

entity mux_circuit is
  port( x, y, z: in std_logic;
        f: out std_logic );
end entity;

architecture logic_flow of mux_circuit is
begin
        f<=(x and not y) or (x and not z) or (y and z);
end architecture;
```

PR 2.14 Program 2.14

Using the **when** VHDL statement, we can implement the circuit in Fig. 2.3 as in PR 2.15.

PR 2.15 Program 2.15

```
library ieee;
use ieee.std_logic_1164.all;

entity mux_circuit is
  port( x, y, z: in std_logic;
        f: out std_logic );
end entity;

architecture logic_flow of mux_circuit is
begin
    f<=x when ((y and z)='0') else
        y ;
end architecture;
```

Using the **select** VHDL statement, we can implement the circuit in Fig. 2.3 as in PR 2.16.

PR 2.16 Program 2.16

```
library ieee;
use ieee.std_logic_1164.all;

entity mux_circuit is
  port( x, y, z: in std_logic;
        f: out std_logic );
end entity;

architecture logic_flow of mux_circuit is

signal yz: std_logic_vector(1 downto 0);

begin
  yz<=y&z;
  with (yz) select
      f<=x when "00",
         x when "01",
         x when "10",
         y when others;
end architecture:
```

Example 2.15 Implement the digital circuit shown in Fig. 2.4 in VHDL. Use **when** statement in your implementation.

Fig. 2.4 Digital circuit for Example 2.15

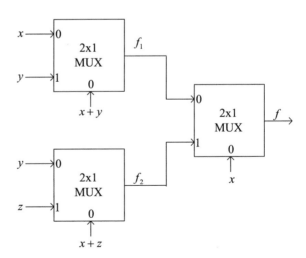

Solution 2.15 First we write the entity part of the program and define the port identities as in PR 2.17.

PR 2.17 Program 2.17

```
library ieee;
use ieee.std_logic_1164.all;

entity mux_circuit is
  port( x, y, z: in std_logic;
        f: out std_logic );
end entity;
```

Next, the internal signals f_1 and f_2 are defined in the declarative part of the architecture unit as in PR 2.18.

PR 2.18 Program 2.18

```
library ieee;
use ieee.std_logic_1164.all;

entity mux_circuit is
  port( x, y, z: in std_logic;
        f: out std_logic );
end entity;

architecture logic_flow of mux_circuit is
  signal f1: std_logic;
  signal f2: std_logic;

begin

end architecture;
```

The architecture part can be written in a number of different ways. The architecture part using the **when** statement is implemented in PR 2.19.

PR 2.19 Program 2.19

```
library ieee;
use ieee.std_logic_1164.all;

entity mux_circuit is
  port( x, y, z: in std_logic;
          f: out std_logic );
end entity;

architecture logic_flow of mux_circuit is
  signal f1: std_logic;
  signal f2: std_logic;

begin
    f1<=x when ((x or y)='0') else
        y ;
    f2<=y when ((x or z)='0') else
        z;
    f<=f1 when (x='0') else
        f2;
end architecture;
```

Note that the sequence orders of the statements in the architecture part of PR 2.19 is not important. Since all the statements in the architecture part are concurrently run. The architecture part of PR 2.19 can also be written as in PR 2.20.

PR 2.20 Program 2.20

```
architecture logic_flow of mux_circuit is

  signal f2: std_logic;
  signal f1: std_logic;

begin
    f2<=y when ((x or z)='0') else
        z;
    f<=f1 when (x='0') else
        f2;
    f1<=x when ((x or y)='0') else
        y ;
end architecture;
```

Example 2.16 Implement the digital circuit shown in Fig. 2.5 in VHDL. Use **when** statement in your implementation.

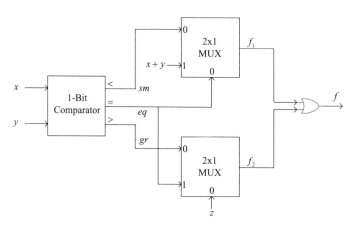

Fig. 2.5 Digital circuit for Example 2.16

Solution 2.16 First, we define input output port parameters x, y, f in the **entity** part as in PR 2.21.

PR 2.21 Program 2.21

```
library ieee;
use ieee.std_logic_1164.all;

entity comb_circuit is
  port( x, y, z: in std_logic;
        f: out std_logic );
end entity;
```

Second, we define the internal signals sm, eq, gr, f_1 and f_2 in the declarative part of the **architecture** unit as in PR 2.22.

```
library ieee;
use ieee.std_logic_1164.all;

entity comb_circuit is
  port( x, y, z:in std_logic;
            f: out std_logic);
end entity;

architecture logic_flow of comb_circuit is
  signal f1, f2: std_logic;
  signal sm, eq, gr: std_logic;
begin
```

PR 2.22 Program 2.22

In the final part, the body part of the architecture unit is written considering the operation of comparator and multiplexer units of the circuit in Fig. 2.5 as in PR 2.23.

PR 2.23 Program 2.23

```
library ieee;
use ieee.std_logic_1164.all;

entity comb_circuit is
  port( x, y, z: in std_logic;
            f: out std_logic );
end entity;

architecture logic_flow of comb_circuit is
  signal f1, f2: std_logic;
  signal sm, eq, gr: std_logic;

begin
-- Bit Comparator Implementation
  sm <= '1' when (x<y) else
             '0';
  eq <= '1' when (x=y) else
             '0';
  gr <= '1' when (x>y) else
             '0';
--- MUX Implementation
  f1 <= sm when (eq='0') else -- MUX1
             x or y;
  f2 <= gr when (z='0') else -- MUX2
             eq;
  f <= f1 or f2; -- OR gate output
end architecture:
```

2 × 1 Multiplexer with bus inputs and outputs

Multiplexer inputs may accept logic vectors. The black box representation of a
2 × 1 multiplexer accepting 8-bit logic vectors is shown in Fig. 2.6.

Fig. 2.6 2 × 1 multiplexer
with 8-bit busses

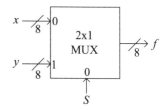

Example 2.17 In PR 2.24 and 2.25, implementation of 2 × 1 multiplexers
accepting logic vectors are given.

PR 2.24 Program 2.24

```
library ieee;
use ieee.std_logic_1164.all;

entity multiplexer_2x1 is
  port( x, y: in std_logic_vector(7 downto 0);
        s: in std_logic;
        f: out std_logic_vector(7 downto 0));
end entity;

architecture logic_flow of multiplexer_2x1 is

begin
        f<=x when s='0' else
           y ;
end architecture;
```

PR 2.25 Program 2.25

```
library ieee;
use ieee.std_logic_1164.all;

entity multiplexer_2x1 is
  port( x, y: integer;
        s: in integer range 0 to 1;
        f: out integer);
end entity;

architecture logic_flow of multiplexer_2x1 is

begin
        f<=x when s=0 else
            y ;
end architecture;
```

In PR 2.25, in the **port** declarations, for the ports x and y, direction of the ports is not indicated. In this case, default direction, i.e., **input**, is accepted. And for the ports x, y and f the data type is integer, but, a range is not defined for the integer data type. In this case, the default range is accepted. The default range for the integers is the same as 32-bit signed integer range, i.e., from -2^{31} to $2^{31} - 1$.

4×1 **Multiplexer**

The black-box representation of 4×1 multiplexer is shown in Fig. 2.7.

Fig. 2.7 4×1 multiplexer

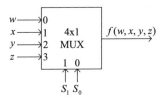

Table 2.4 Truth table of 4×1 multiplexer

S_1	S_0	f		w	x	y	z	S_1	S_0	f
0	0	f_0	$\leftrightarrow$	0	0	0	0	0	0	0
0	1	f_1		0	0	0	1	0	0	0
1	0	f_2		0	0	1	0	0	0	0
1	1	f_3		0	0	1	1	0	0	0
				0	1	0	0	0	1	1
				0	1	0	1	0	1	1
				0	1	1	0	0	1	1
				0	1	1	1	0	1	1
				0	0	0	0	1	0	0
				1	0	0	1	1	0	0
				1	0	1	0	1	0	1
				1	0	1	1	1	0	1
				1	1	0	0	1	1	0
				1	1	0	1	1	1	1
				1	1	1	0	1	1	0
				1	1	1	1	1	1	1

The truth table of the 4×1 multiplexer is depicted in Table 2.4 in two different ways.

For a 4×1 multiplexer, the internal circuit diagram is depicted in Fig. 2.8.

Fig. 2.8 Internal circuit for 4×1 multiplexer

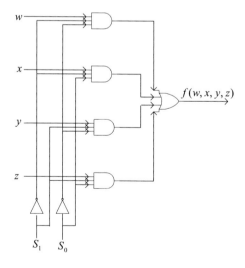

Example 2.18 Implement the 4×1 multiplexer shown in Fig. 2.7 in VHDL language.

Solution 2.18 The implementation of 4×1 multiplexer is given in PR 2.26.

```
library ieee;
use ieee.std_logic_1164.all;

entity multiplexer_4x1 is
  port( w, x, y, z, s1, s0: in std_logic;
        f: out std_logic );
end entity;

architecture logic_flow of multiplexer_4x1 is
begin
        f<=w when (s1='0' and s0='0') else
```

```
x when (s1='0' and s0='1') else
y when (s1='1' and s0='0') else
z;

end architecture;
```

PR 2.26 Program 2.26

Example 2.19 The 4×1 multiplexer shown in Fig. 2.7 in can alternatively be implemented as in PR 2.27, 2.28, 2.29.

```
library ieee;
use ieee.std_logic_1164.all;

entity multiplexer_4x1 is
  port( w, x, y, z: in std_logic;
        s1s0: in std_logic_vector(1 downto 0);
        f: out std_logic );
end entity;
```

```
architecture logic_flow of
multiplexer_4x1 is
begin

f<=w when (s1s0="00") else
   x when (s1s0="01") else
   y when (s1s0="10") else
   z;

end architecture;
```

PR 2.27 Program 2.27

```
library ieee;
use ieee.std_logic_1164.all;

entity multiplexer_4x1 is
  port( w, x, y, z: in std_logic;
        s1s0: in natural range 0 to 3;
        f: out std_logic );
end entity;
```

```
architecture logic_flow of
multiplexer_4x1 is
begin

f<=w when (s1s0=0) else
   x when (s1s0=1) else
   y when (s1s0=2) else
   z;

end architecture;
```

PR 2.28 Program 2.28

```
library ieee;
use ieee.std_logic_1164.all;

entity multiplexer_4x1 is
  port( wxyz: in std_logic_vector(3 downto 0);
        s1s0: in natural range 0 to 3;
        f: out std_logic );
end entity;
```

```
architecture logic_flow of
multiplexer_4x1 is
begin

  f<= wxyz(3) when (s1s0=0) else
      wxyz(2) when (s1s0=1) else
      wxyz(1) when (s1s0=2) else
      wxyz(0);

end architecture;
```

PR 2.29 Program 2.29

Example 2.20 Implement the 4×1 multiplexer shown in Fig. 2.7 in VHDL language. Use **select** statement for your implementation.

Solution 2.20 The VHDL implementation of 4×1 multiplexer with **select** statement is depicted in PR 2.30.

```
library ieee;
use ieee.std_logic_1164.all;

entity multiplexer_4x1 is
  port( w, x, y, z: in std_logic;
        s1s0: in std_logic_vector(1 downto 0);
        f: out std_logic );
end entity;
```

```
architecture logic_flow of
multiplexer_4x1 is
begin

with s1s0 select
f<=w when ("00"),
   x when ("01"),
   y when ("10")
   z others;

end architecture;
```

PR 2.30 Program 2.30

Note that the use of **others** in **with—select** statement is a must, if all the conditions are not tested, for instance, we did not include "ZZ", "WZ", etc., in conditional expression, and in this case, **others** should be included in the last conditional expression.

Example 2.21 Implement the 4×1 multiplexer whose inputs are 9-bit signed integers.

Solution 2.21 The VHDL implementation to 4×1 multiplexer with 9-bit signed integers is depicted in PR 2.31.

```
library ieee;
use ieee.numeric_std.all;

entity multiplexer_4x1 is
  port( w, x, y, z: signed(8 downto 0);
        s1s0: integer range 0 to 3;
        f: out signed(8 downto 0) );
end entity;
```

```
architecture logic_flow of multiplexer_4x1
is
begin

f<=w when (s1s0=0) else
   x when (s1s0=1) else
   y when (s1s0=2) else
   z;

end architecture;
```

PR 2.31   Program 2.31

Example 2.22 Convert the conditional **when** statement in PR 2.32 into a conditional **select** statement.

PR 2.32   Program 2.32

```
f<=w when (s1s0=0) else
   x when (s1s0=1) else
   y when (s1s0=2) else
   z;
```

Solution 2.22 The conversion is outlined as follows.
  (S1) First add **with…select** to the top as in PR 2.33.

PR 2.33   Program 2.33

```
with…select
f<=w when (s1s0=0) else
   x when (s1s0=1) else
   y when (s1s0=2) else
   z;
```

  (S2) Move the conditional parameter of **when** statement into the conditional parameter of **select** statement, and leave the numeric values of the conditional parameters of **when** as in PR 2.34.

PR 2.34   Program 2.34

```
with s1s0 select
f<=w when (0) else
   x when (1) else
   y when (2) else
   z;
```

(S3) Remove reserved word **else** and put a comma after each line except the last one as in PR 2.35.

PR 2.35 Program 2.35

```
with s1s0 select
  f<=w when (0),
      x when (1),
      y when (2),
      z;
```

(S4) Add the word **when others** to the last line as in PR 2.36.

PR 2.36 Program 2.36

```
with s1s0 select
  f<=w when (0),
      x when (1),
      y when (2),
      z when others;
```

Example 2.23 Implement the circuit shown in Fig. 2.9 in VHDL language. Use **when** statement in your implementation.

Fig. 2.9 Digital circuit for Example 2.23

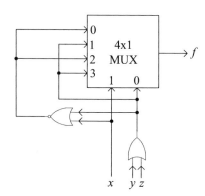

Solution 2.23 Let f_0 and f_1 be the outputs of the OR and NOR gates respectively, and I_0, I_1, I_2, I_3 are the inputs of the 4 × 1 multiplexer as shown in Fig. 2.10. Note that f_0, f_1, I_0, I_1, I_2 and I_3 are nothing but internal signals of the circuit.

Fig. 2.10 Digital circuit with internal signals labeled

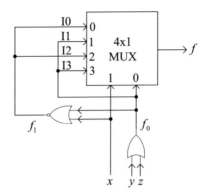

We define I/O signals of the circuit at the **entity** part, and define the internal signals at the declarative part of the architecture as in PR 2.37.

PR 2.37 Program 2.37

```
library ieee;
use ieee.std_logic_1164.all;

entity multiplexer_4x1 is
    port( w, x, y, z: in std_logic;
            f: out std_logic );
end entity;

architecture logic_flow of multiplexer_4x1 is
    signal f0, f1, I0, I1, I2, I3: std_logic;
begin

end architecture;
```

Considering Fig. 2.10, we can write the functions f_0, f_1, I_0, I_1, I_2 and I_3 as in

$$f_0 = y + z \quad f_1 = (x + f_0)' \quad I_0 = f_1 \quad I_1 = f_0 \quad I_2 = f_1 \quad I_3 = f_0. \quad (2.8)$$

The implementation of the internals signals and circuit output can be written in the architecture part as indicated in PR 2.38.

PR 2.38 Program 2.38

```
library ieee;
use ieee.std_logic_1164.all;

entity multiplexer_4x1 is
  port( w, x, y, z: in std_logic;
          f: out std_logic );
end entity;

architecture logic_flow of multiplexer_4x1 is

  signal f0, f1, I0, I1, I2, I3: std_logic;

begin
  f0<= y or z;
  f1<= x nor (y or z);
  I0<=f1;
  I1<=f0;
  I2<=f1;
  I3<=f0;
   f<=I0 when (x= '0' and f0='0') else
      I1 when (x= '0' and f0='1') else
      I2 when (x= '1' and f0='0') else
      I3;
end architecture;
```

Example 2.24 Implement the circuit shown in Fig. 2.10 in VHDL language. Use **select** statement in your implementation.

Solution 2.24 We can just convert the **when** statement of the previous solution as in PR 2.39.

PR 2.39 Program 2.39

```
S=x&f0;
with S select
   f<=I0 when ("00"),
      I1 when ("01"),
      I2 when ("10"),
      I3 when others;
```

Gray Code

Grad code is used to encode the integers $0, 1, 2, \ldots, 15$. The binary strings used for encoding of consecutive numbers differ in only 1-bit location. The Gray coding is illustrated in Table 2.5.

Table 2.5 Gray code table

Number	Gray code
0	0000
1	0001
2	0011
3	0010
4	0110
5	0111
6	0101
7	0100
8	1100
9	1101
10	1111
11	1110
12	1010
13	1011
14	1001
15	1000

Example 2.25 Implement the Gray coding in VHDL language.

Solution 2.25 The Gray coding operation illustrated in Table 2.3 can be implemented as in PR 2.40.

```
library ieee;
use ieee.std_logic_1164.all;

entity gray_coding is
  port( number: in natural range 0 to 15;
        codeword: out std_logic_vector(3 downto 0) );
end entity;

architecture logic_flow of gray_coding is
begin
  codeword<= "0000" when number=0 else
             "0001" when number=1 else
             "0011" when number=2 else
             "0010" when number=3 else
             "0110" when number=4 else
             "0111" when number=5 else
             "0101" when number=6 else
             "0100" when number=7 else
             "1100" when number=8 else
             "1101" when number=9 else
             "1111" when number=10 else
             "1110" when number=11 else
             "1010" when number=12 else
             "1011" when number=13 else
             "1001" when number=14 else
             "1000";

end architecture;
```

PR 2.40 Program 2.40

Exercise Implement the BCD code illustrated in Table 2.6.

Table 2.6 BCD code table

0 ↔ 0000
1 ↔ 0001
2 ↔ 0010
3 ↔ 0011
4 ↔ 0100
5 ↔ 0101
6 ↔ 0110
7 ↔ 0111
8 ↔ 1000
9 ↔ 1001

Octal Encoder: Encoders are employed to represent integers by binary strings. Binary encoders are used to represent integers by binary numbers expressed in base two. One of the binary encoders is the octal encoder. It is used to represent the integers $0, 1, 2, \ldots, 7$. It accepts a digit in the range 0–7 and outputs a 3-bit binary string. The black box representation of octal encoder is depicted in Fig. 2.11.

Fig. 2.11 Octal encoder black box representation

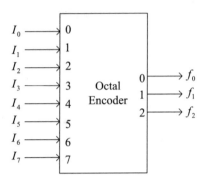

Considering the operation of the octal encoder, we can construct its truth table as in Table 2.7.

Table 2.7 Octal encoder truth table

I_7	I_6	I_5	I_4	I_3	I_2	I_1	I_0	f_2	f_1	f_0
0	0	0	0	0	0	0	1	0	0	0
0	0	0	0	0	0	1	0	0	0	1
0	0	0	0	0	1	0	0	0	1	0
0	0	0	0	1	0	0	0	0	1	1
0	0	0	1	0	0	0	0	1	0	0
0	0	1	0	0	0	0	0	1	0	1
0	1	0	0	0	0	0	0	1	1	0
1	0	0	0	0	0	0	0	1	1	1

Example 2.26 Implement the octal encoding operation in VHDL language.

Solution 2.26 The octal encoding operation illustrated in Table 2.5 can be implemented as in PR 2.41, 2.42 and 2.43. In PR 2.41, the implementation is performed using **with-select** statement. On the other hand, in PR 2.42 and 2.43, the implementation is done using the **when** statement.

```
library ieee;
use ieee.std_logic_1164.all;

entity octal_encoder is
  port( I:in std_logic_vector(7 downto 0);
          F: out std_logic_vector(2 downto 0) );
end entity;

architecture logic_flow of octal_encoder is
begin
with (I) select
  F<="000"when "00000001",
      "001"when "00000010",
      "010"when "00000100",
      "011"when "00001000",
      "100"when "00010000",
      "101"when "00100000",
      "110"when "01000000",
      "111"when others;
end architecture;
```

PR 2.41 Program 2.41

```vhdl
library ieee;
use ieee.std_logic_1164.all;

entity octal_encoder is
 port( I: in std_logic_vector(7 downto 0);
       F: out std_logic_vector(2 downto 0) );
end entity;

architecture logic_flow of octal_encoder is
begin

 F<="000" when I="00000001" else
     "001" when I="00000010" else
     "010" when I="00000100" else
     "011" when I="00001000" else
     "100" when I="00010000" else
     "101" when I="00100000" else
     "110" when I="01000000" else
     "111";
end architecture;
```

PR 2.42 Program 2.42

```vhdl
entity octal_encoder is
 port( I: in positive range 128 downto 1;
       F: out natural range 7 downto 0);
end entity;

architecture logic_flow of octal_encoder is
begin

 F<=0 when I=1  else
    1 when I=2  else
    2 when I=4  else
    3 when I=8  else
    4 when I=16 else
    5 when I=32 else
    6 when I=64 else
    7;
end architecture;
```

PR 2.43 Program 2.43

Priority Encoder

In priority encoder, the integers to be encoded have different priorities. If a numbers of integers are to be encoded at the same time, then the number having the largest priority is encoded. Such encoders are called priority encoders.

The black box representation of a typical octal priority encoder is depicted in Fig. 2.12.

Fig. 2.12 Octal priority encoder black box representation

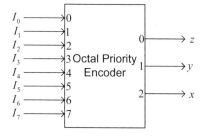

The operation of the octal priority encoder is illustrated in Fig. 2.13. Let's assume that the integers 1, 3 and 5 are to be encoded at the same time. The integer 5 which has the highest priority is encoded as illustrated in Fig. 2.13.

Fig. 2.13 Operation of the octal priority encoder

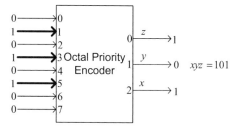

The black box representation of a 4-bit priority encoder with valid output indicator bit, i.e., bit v, is depicted in Fig. 2.14.

Fig. 2.14 4-bit priority encoder black box representation

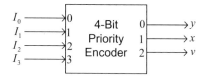

The truth table of the 4-bit priority encoder with valid result indicator is shown Table 2.8.

Table 2.8 Truth table of 4-bit priority encoder

I_3	I_2	I_1	I_0	x	y	z
1	×	×	×	1	1	1
0	1	×	×	1	0	1
0	0	1	×	0	1	1
0	0	0	1	0	0	1
0	0	0	0	×	×	0

Example 2.27 Implement the 4-bit priority encoder, whose truth table is shown Table 2.8, in VHDL language.

Solution 2.27 The priority encoding operation illustrated in Table 2.8 can be implemented as in PR 2.44. For the don't care condition, we use '−' symbol, for instance "1 × ××" can be implemented in VHDL as "1 − −−".

```
library ieee;
use ieee.std_logic_1164.all;

entity Four_bit_priority_encoder is
  port( I: in std_logic_vector(3 downto 0);
        xyv: out std_logic_vector(2 downto 0) );
end entity;

architecture logic_flow of Four_bit_priority_encoder is
begin
with (I) select
  xyv<="111" when "1---",
       "101" when "01--",
       "011" when "001-",
       "001" when "0001",
       "--0" when others;
end architecture;
```

PR 2.44 Program 2.44

Binary Coded Decimal (BCD) Encoding

BCD encoding is used to encode only the digits, $0, 1, 2, \ldots, 9$. It is not used to encode the numbers greater than 9. The BCD encoding logic is illustrated in Table 2.9.

Table 2.9 BCD encoding operation

0 ↔ 0000	
1 ↔ 0001	
2 ↔ 0010	
3 ↔ 0011	
4 ↔ 0100	
5 ↔ 0101	
6 ↔ 0110	
7 ↔ 0111	
8 ↔ 1000	
9 ↔ 1001	

Decoders

The task of the decoders is the opposite of that of the encoders. The codeword, i.e., binary string, is resolved to the character it represents. The encoder and decoder units are designed in a coherent manner to each other. One of the decoders used in digital design is the BCD decoder whose black box representation is depicted in Fig. 2.15.

Fig. 2.15 BCD decoder black box representation

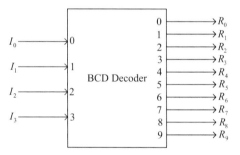

The operation of the BCD decoder is illustrated in Fig. 2.16 where at the input of the decoder we have a codeword, and at the output of the decoder the integer represented by the codeword is indicated by the logic value '1'.

Fig. 2.16 Operation of BCD decoder

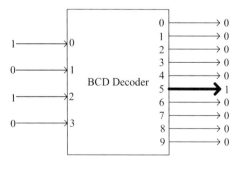

As it is seen from Fig. 2.16, the codeword '0101' when resolved activates the output port indexed by 5. That means that the codeword '0101' represent the integer 5.

Example 2.28 Implement the BCD decoder in VHDL language.

Solution 2.28 Referring to Table 2.9, BCD decoder can be implemented in VHDL language as in PR 2.45.

PR 2.45 Program 2.45

```
library ieee;
use ieee.std_logic_1164.all;

entity BCD_decoder is
  port( I: in std_logic_vector(3 downto 0);
          F: out natural range 1 to 512 );
end entity;

architecture logic_flow of BCD_decoder is
begin
  F<=1 when I="0000" else
      2 when I="0001" else
      4 when I="0010" else
      8 when I="0011" else
     16 when I="0100" else
     32 when I="0101" else
     64 when I="0110" else
    128 when I="0111" else
    256 when I="1000" else
    512;
end architecture;
```

BCD-to-Seven Segment Display Converter

The seven segment display an electronic unit is used to display the digits 0, 1, ..., 9. The symbolic representation of seven segment display is depicted in Fig. 2.17.

Fig. 2.17 Seven segment display symbolic representation

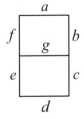

The BCD codes representing digits and their corresponding seven segment display codes are depicted in Table 2.10.

Table 2.10 BCD and SS display codes

Digit	BCD code	Seven segment display code
	wxyz	*abcdefg*
0	0000	0000001
1	0001	1001111
2	0010	0010010
3	0011	0000110
4	0100	1001100
5	0101	0100100
6	0110	0100000
7	0111	0001111
8	1000	0000000
9	1001	0000100

Example 2.29 Implement the BCD to Seven Segment Display Converter in VHDL language.

Solution 2.29 Referring to Table 2.10, BCD to Seven Segment Display Converter in VHDL language can be implemented as in PR 2.46.

```
library ieee;
use ieee.std_logic_1164.all;

entity BCD_to_SS_converter is
  port( BCD: in std_logic_vector(3 downto 0);
        SSD: out std_logic_vector(6 downto 0) );
end entity;

architecture logic_flow of BCD_to_SS_converter is
begin
  SSD<="0000001" when BCD="0000" else
       "1001111" when BCD="0001" else
       "0010010" when BCD="0010" else
       "0000110" when BCD="0011" else
       "1001100" when BCD="0100" else
       "0100100" when BCD="0101" else
       "0100000" when BCD="0110" else
       "0001111" when BCD="0111" else
       "0000000" when BCD="1000" else
       "0000100";
end architecture;
```

PR 2.46 Program 2.46

Problems

(1) Implement the Boolean function $f = x'y + xz + y'z$ in VHDL.
(2) Implement the hexadecimal binary encoder in VHDL.
(3) Implement the hexadecimal binary decoder in VHDL.
(4) Implement the truth table of the Boolean function given in Table 2.P4 in VHDL.

Table 2.P4 Truth table of a Boolean function

xyz	$f(x,y,z)$
000	1
001	1
010	0
011	1
100	1
101	0
110	1
111	1

(5) Implement the combinational circuit shown in Fig. 2.P5 in VHDL.

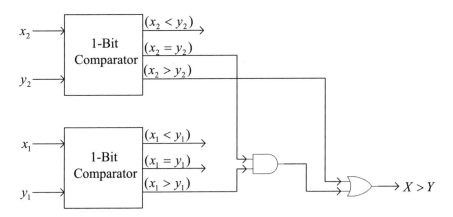

Fig. 2.P5 A combinational circuit diagram

(6) Implement the combinational circuit shown in Fig. 2.P6 in VHDL.

Fig. 2.P6 Combinational
circuit with a multiplexer

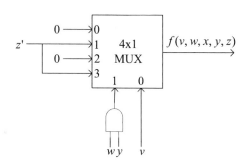

(7) Implement the combinational circuit shown in Fig. 2.P7 in VHDL.

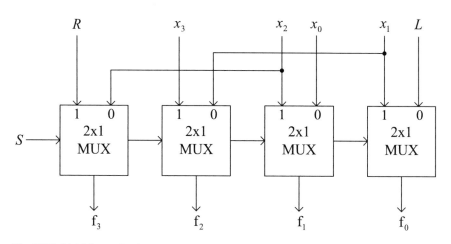

Fig. 2.P7 Multiplexer circuit

(8) Implement the combinational circuit shown in Fig. 2.P8 in VHDL.

Fig. 2.P8 Combinational circuit involving multiplexer and de-multiplexer

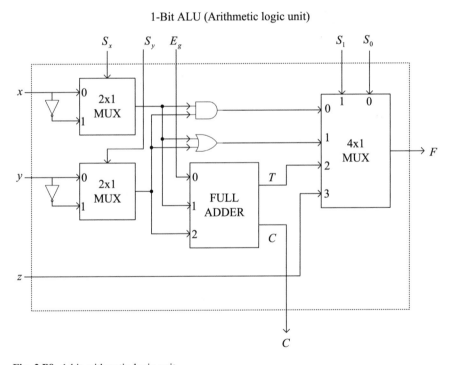

(9) Implement the combinational circuit shown in Fig. 2.P9 in VHDL

1-Bit ALU (Arithmetic logic unit)

Fig. 2.P9 1-bit arithmetic logic unit

Chapter 3
Simulation of VHDL Programs

In this chapter we will give information about the simulation of VHDL programs. Once we implement a digital circuit or an algorithm in VHDL language, we need to check the correctness of the implementation. For this purpose, we write test programs which are used to supply values to the input ports of the VHDL program, and it is possible to observe the values at the output ports. The test program written to simulate the input and output behavior of the VHDL programs is usually called test-bench.

If we cannot observe the expected output values for the supplied input values, we decide that the written VHDL program has some bugs inside, and we check the written program for possible mistakes. We repeat this procedure until we observe the expected values at output ports for the supplied input values.

3.1 Test-Bench Writing

The best way to learn how to write a test-bench passes through an example. Let's illustrate how to write a test-bench with the following example.

Example 3.1 Write a test-bench for the simulation of the Boolean function $f(x, y, z) = x'y' + y'z$ implemented in VHDL.

Solution 3.1 First, we implement the Boolean function as in PR 3.1.

© Springer Nature Singapore Pte Ltd. 2019
O. Gazi, *A Tutorial Introduction to VHDL Programming*,
https://doi.org/10.1007/978-981-13-2309-6_3

To test the VHDL program given in PR 3.1, we can write a test-bench considering the following steps.

PR 3.1 Program 3.1

```
library ieee;
use ieee.std_logic_1164.all;

entity f_xyz is
  port( x, y, z: in std_logic;
           f: out std_logic );
end entity;

architecture logic_flow of f_xyz is

begin
        f<= (x nor y) or (not y and z);
end architecture;
```

(S1) In Step-1, we write an empty **entity** part as in PR 3.2. The name of the **entity** for the test-bench can be chosen anything, however, the convention is to add the suffix **_TB** to the end of the entity name of the VHDL unit to be tested, i.e., for this example it is formed as f_xyz_TB.

PR 3.2 Program 3.2

```
library ieee;
use ieee.std_logic_1164.all;

entity f_xyz_TB is
end entity;
```

(S2) In Step-2, we add the structure of the **architecture** unit as in PR 3.3 where declarative part, the part between **architecture** and **begin**, and the body part, the part between **begin** and **end architecture,** of the **architecture** unit are displayed empty.

PR 3.3 Program 3.3

```
library ieee;
use ieee.std_logic_1164.all;

entity f_xyz_TB is
end entity;

architecture logic_flow of f_xyz_TB is


begin



end architecture;
```

(S3) In Step-3, we copy the **entity** unit of the main program to be tested to the declarative part of the architecture unit as in PR 3.4.

PR 3.4 Program 3.4

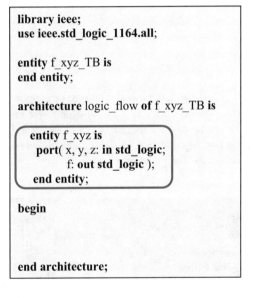

```
library ieee;
use ieee.std_logic_1164.all;

entity f_xyz_TB is
end entity;

architecture logic_flow of f_xyz_TB is

    entity f_xyz is
      port( x, y, z: in std_logic;
            f: out std_logic );
    end entity;

begin



end architecture;
```

(S4) In Step-4, in the copied **entity** unit of PR 3.4, change the word **entity** to **component**, and drop the word **is** as in PR 3.5. Note that it is optional to drop the word "**is**" from the end of the component name, it can be kept as well.

PR 3.5 Program 3.5

```
library ieee;
use ieee.std_logic_1164.all;

entity f_xyz_TB is
end entity;

architecture logic_flow of f_xyz_TB is

    component f_xyz
      port( x, y, z: in std_logic;
              f: out std_logic );
    end component;

begin



end architecture;
```

(S5) In Step-5, the initial values to be supplied to the input ports are placed below the **component** declaration in the declarative part of the architecture unit as in PR 3.6. The name of the parameters to supply the initial input port values can be the same as the port input names, or they can be chosen differently. In PR 3.6 we used port input names for the parameter names.

```
library ieee;
use ieee.std_logic_1164.all;

entity f_xyz_TB is
end entity;

architecture logic_flow of f_xyz_TB is

    component f_xyz
      port( x, y, z: in std_logic;
              f: out std_logic );
    end component;
```

```
    signal x: std_logic:='0';
    signal y: std_logic:='0';
    signal z: std_logic:='0';
    signal f: std_logic;

begin


end architecture;
```

PR 3.6 Program 3.6

(S6) In Step-6, Component instantiation is done in the body part of the **architecture** unit as seen in PR 3.7, and the initial values to be supplied to the input ports are send to the input ports using the **port map** statement as in PR 3.7.

```
library ieee;
use ieee.std_logic_1164.all;

entity f_xyz_TB is
end entity;

architecture logic_flow of f_xyz_TB is

    component f_xyz
        port( x, y, z: in std_logic;
              f: out std_logic );;
    end component;
```

```
signal x: std_logic:='0';
signal y: std_logic:='0';
signal z: std_logic:='0';
signal f: std_logic;

begin

    pm: f_xyz port map(
        x=>x,
        y=>y,
        z=>z ,
        f=>f);
```

PR 3.7 Program 3.7

If we had used different names for the initialization parameters other than the same port names, then we would write our program as in PR 3.8.

```
library ieee;
use ieee.std_logic_1164.all;

entity f_xyz_TB is
end entity;

architecture logic_flow of f_xyz_TB is

    component f_xyz
        port( x, y, z: in std_logic;
              f: out std_logic );;
    end component;
```

```
signal x1: std_logic:='0';
signal y1: std_logic:='0';
signal z1: std_logic:='0';
signal f1: std_logic;

begin

pm: f_xyz port map(
    x => x1,
    y => y1,
    z => z1,
    f=>f1);
```

PR 3.8 Program 3.8

Note that in the assignments $y <= x$ and $w => z$, the values are at the right and the destination parameters are on the left hand side.

(S7) After initial value supplement stage, in Step-7, we provide different input values to the input ports. For this purpose, we use **process** units of the VHDL language. We can write more than one **process** units to provide values to the input ports. Keep in your mind that **process** units are sequential program units, and if more than one **process** is written, they are run in parallel, although they are sequential inside. In PR 3.9, we used a single process unit with simulation values.

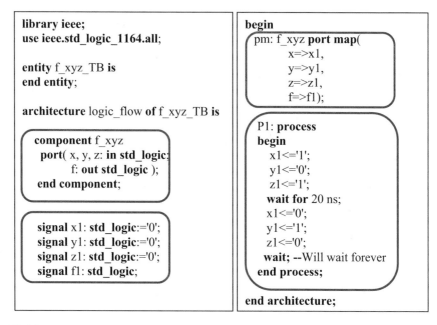

library ieee;
use ieee.std_logic_1164.all;

entity f_xyz_TB **is**
end entity;

architecture logic_flow **of** f_xyz_TB **is**

  component f_xyz
    port(x, y, z: **in std_logic**;
        f: **out std_logic**);
  end component;

  signal x1: **std_logic**:='0';
  signal y1: **std_logic**:='0';
  signal z1: **std_logic**:='0';
  signal f1: **std_logic**;

begin
  pm: f_xyz **port map**(
        x=>x1,
        y=>y1,
        z=>z1,
        f=>f1);

  P1: **process**
  begin
    x1<='1';
    y1<='0';
    z1<='1';
    wait for 20 ns;
    x1<='0';
    y1<='1';
    z1<='0';
    wait; --Will wait forever
  end process;

end architecture;

PR 3.9 Program 3.9

The **wait** statement at the end of process unit guarantees the simulation to be performed once for the supplied values. If **wait** statement is not placed at the end of process unit, then the simulation is repeated over and over again with the same simulation values, and the simulation graph shows the simulation results in a periodic manner, i.e., the same graph is repeated in time.

We can use three **process** units to provide the values for the input ports as in PR 3.10. In fact, we can use any number of **process** units.

The test-bench shown in PR 3.10 can be used for simulation purposes. For illustration purposes, we used the XILINX ISE platform for the simulation process. In XILINX ISE platform, it is possible to see the simulation results as time waveforms. For our test-bench, the simulation results are expressed as time waveforms in Fig. 3.1 where the grey areas indicate the logic '1'.

```
library ieee;
use ieee.std_logic_1164.all;

entity f_xyz_TB is
end entity;

architecture logic_flow of f_xyz_TB is

    component f_xyz
     port( x, y, z: in std_logic;
            f: out std_logic );
    end component;


    signal x1: std_logic:='0';
    signal y1: std_logic:='0';
    signal z1: std_logic:='0';
    signal f1: std_logic;

begin

pm: f_xyz port map(
        x=>x1,
        y=>y1,
        z=>z1,
        f=>f1);

p1: process
    begin
      x1<='0';
      wait for 400 ns;
      x1<='1';
      wait for 400 ns;
      wait;
    end process;
```

```
p2: process
   begin
      y1<='0';
      wait for 200 ns;
      y1<='1';
      wait for 200 ns;
      y1<='0';
      wait for 200 ns;
      y1<='1';
      wait for 200 ns;
      wait;
   end process;

p3: process
   begin
      z1<='0';
      wait for 100 ns;
      z1<='1';
      wait for 100 ns;
      z1<='0';
      wait for 100 ns;
      z1<='1';
      wait for 100 ns;
      z1<='0';
      wait for 100 ns;
      z1<='1';
      wait for 100 ns;
      z1<='0';
      wait for 100 ns;
      z1<='1';
      wait for 100 ns;
      wait;
   end process;
end architecture;
```

PR 3.10 Program 3.10

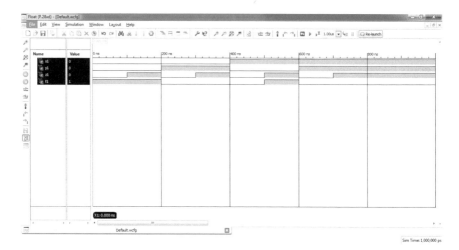

Fig. 3.1 Test-bench simulation results in XILINX ISE platform

The truth table of the Boolean function $f(x, y, z) = x'y' + y'z$ is shown in Table 3.1.

Table 3.1 Truth table of $f = x'y' + y'z$

x	y	z	$f(x, y, z)$
0	0	0	1
0	0	1	1
0	1	0	0
0	1	1	0
1	0	0	0
1	0	1	1
1	1	0	0
1	1	1	0

If Table 3.1 is compared to the simulation parameter values in PR 3.10, it is seen that the simulation values in PR 3.10 are nothing but the values taken from the truth table converted to time waveforms using the **wait for** statement.

The test-bench in PR 3.10 can also be written as in PR 3.11 where it is seen that we use the same input port names for the names of the initialization parameters.

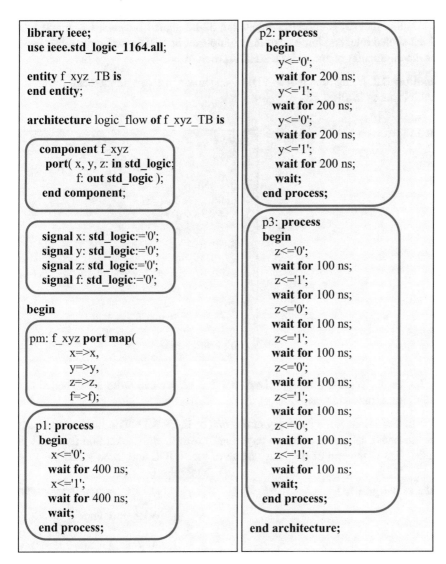

```
library ieee;
use ieee.std_logic_1164.all;

entity f_xyz_TB is
end entity;

architecture logic_flow of f_xyz_TB is

    component f_xyz
      port( x, y, z: in std_logic;
            f: out std_logic );
    end component;

    signal x: std_logic:='0';
    signal y: std_logic:='0';
    signal z: std_logic:='0';
    signal f: std_logic:='0';

begin

pm: f_xyz port map(
        x=>x,
        y=>y,
        z=>z,
        f=>f);

p1: process
    begin
      x<='0';
      wait for 400 ns;
      x<='1';
      wait for 400 ns;
      wait;
    end process;
```

```
p2: process
    begin
      y<='0';
      wait for 200 ns;
      y<='1';
      wait for 200 ns;
      y<='0';
      wait for 200 ns;
      y<='1';
      wait for 200 ns;
      wait;
    end process;

p3: process
    begin
      z<='0';
      wait for 100 ns;
      z<='1';
      wait for 100 ns;
      z<='0';
      wait for 100 ns;
      z<='1';
      wait for 100 ns;
      z<='0';
      wait for 100 ns;
      z<='1';
      wait for 100 ns;
      z<='0';
      wait for 100 ns;
      z<='1';
      wait for 100 ns;
      wait;
    end process;

end architecture;
```

PR 3.11   Program 3.11

Example 3.2 The black box representation of a 2×1 multiplexer is depicted in Fig. 3.2.

Fig. 3.2 The black box representation of a 2×1 multiplexer

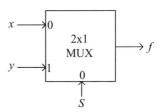

Assume that the multiplexer inputs are 32-bit signed integers, i.e., x and y are 32-bit signed integers. Implement the multiplexer in VHDL, and write a test-bench for the simulation of the written VHDL program.

Solution 3.2 Using the **when** VHDL statement we can implement the 2×1 multiplexer as in PR 3.12.

PR 3.12 Program 3.12

```
library ieee;
use ieee.std_logic_1164.all;

entity multiplexer_2x1 is
  port(x, y: in integer;
           s: in std_logic;
           f: out integer );
end entity;

architecture logic_flow of multiplexer_2x1 is

begin
           f<=x when s='0' else
                 y ;
end architecture;
```

To test the VHDL program given in PR 3.12, we can write a test-bench considering the following steps.

(S1) In Step-1, we write an empty **entity** part as in PR 3.13. The name of the **entity** for the test-bench can be chosen anything, however, the convention is to add the suffix **_TB** to the end of the entity name of the VHDL unit to be tested.

PR 3.13 Program 3.13

```
library ieee;
use ieee.std_logic_1164.all;

entity multiplexer_2x1_TB is
end entity;
```

(S2) In Step-2, we add the structure of the **architecture** unit as in PR 3.14 where declarative part, the part between **architecture** and **begin**, and the body part, the part between **begin** and **end architecture,** of the **architecture** unit are displayed empty.

PR 3.14   Program 3.14

```
library ieee;
use ieee.std_logic_1164.all;

entity multiplexer_2x1_TB is
end entity;

architecture logic_flow of multiplexer_2x1_TB is


begin


end architecture;
```

(S3) In Step-3, we copy the entity unit of the main program to be tested to the declarative part of the test-bench as in PR 3.15.

PR 3.15   Program 3.15

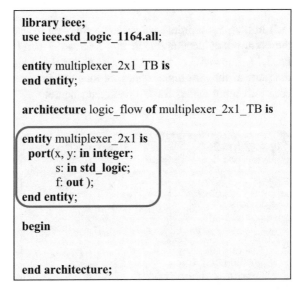

```
library ieee;
use ieee.std_logic_1164.all;

entity multiplexer_2x1_TB is
end entity;

architecture logic_flow of multiplexer_2x1_TB is

entity multiplexer_2x1 is
  port(x, y: in integer;
       s: in std_logic;
       f: out );
end entity;

begin


end architecture;
```

(S4) In Step-4, in the copied **entity** unit of PR 3.15, change the word **entity** to **component**, and we can drop the word **is** as in PR 3.16. Note that dropping the word "**is**" is not a must.

PR 3.16 Program 3.16

```
library ieee;
use ieee.std_logic_1164.all;

entity multiplexer_2x1_TB is
end entity;

architecture logic_flow of multiplexer_2x1_TB is

component multiplexer_2x1
  port(x, y: in integer;
        s: in std_logic;
        f: out integer );
end component;

begin



end architecture;
```

(S5) In Step-5, the initial values to be supplied to the input ports are placed below the **component** declaration in the declarative part of the architecture unit as in PR 3.17. The name of the parameters to supply the initial input port values can be the same as the port input names, or they can be chosen differently. In PR 3.17 we used port input names for the parameter names.

```
library ieee;
use ieee.std_logic_1164.all;

entity multiplexer_2x1_TB is
end entity;

architecture logic_flow of multiplexer_2x1_TB is

component multiplexer_2x1
  port(x, y: in integer;
        s: in std_logic;
        f: out integer );
end component;

signal x: integer:=0;
signal y: integer:=0;
signal s:  std_logic:='0';
signal f:  integer;

begin


end architecture;
```

PR 3.17 Program 3.17

(S6) In Step-6, Component instantiation is done in the body part of the **architecture** unit as seen in PR 3.18, and the initial values to be supplied to the input ports are send to the input ports using the **port map** statement as in PR 3.18.

```
library ieee;                                   signal x: integer:=0;
use ieee.std_logic_1164.all;                    signal y: integer:=0;
                                                signal s:  std_logic:='0';
entity multiplexer_2x1_TB is                    signal f:  integer;
end entity;

architecture logic_flow of multiplexer_2x1_TB is    begin

    component multiplexer_2x1                   pm: multiplexer_2x1 port  map(
    port(x, y: in integer;                          x=>x,
        s: in std_logic;                            y=>y,
        f: out integer );                           s=>s ,
    end component;                                  f=>f);
                                                end architecture;
```

PR 3.18 Program 3.18

If we had used different names for the initialization parameters other than the same port names, then we would write our program as in PR 3.19.

```
library ieee;                                   signal x1: integer:=0;
use ieee.std_logic_1164.all;                    signal y1: integer:=0;
                                                signal s1:  std_logic:='0';
entity multiplexer_2x1_TB is                    signal f1:  integer;
end entity;

architecture logic_flow of multiplexer_2x1_TB is    begin

    component multiplexer_2x1                   pm: multiplexer_2x1 port  map(
    port(x, y: in integer;                          x=>x1,
        s: in std_logic;                            y=>y1,
        f: out integer );                           s=>s1,
    end component;                                  f=>f1);

                                                end architecture;
```

PR 3.19 Program 3.19

(S7) After initial value supplement stage, in Step-7, we provide different input values to the input ports. For this purpose, we use **process** units of the VHDL language. A single process unit with simulation values is shown in PR 3.20.

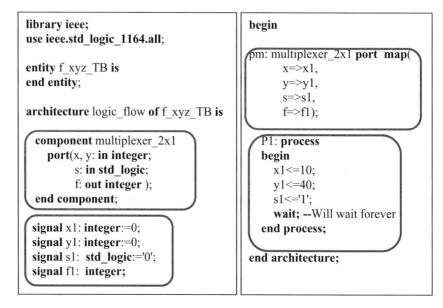

```
library ieee;                                begin
use ieee.std_logic_1164.all;
                                             pm: multiplexer_2x1 port map(
entity f_xyz_TB is                               x=>x1,
end entity;                                      y=>y1,
                                                 s=>s1,
architecture logic_flow of f_xyz_TB is           f=>f1);

   component multiplexer_2x1                  P1: process
      port(x, y: in integer;                  begin
            s: in std_logic;                      x1<=10;
            f: out integer );                     y1<=40;
   end component;                                 s1<='1';
                                                 wait; --Will wait forever
   signal x1: integer:=0;                      end process;
   signal y1: integer:=0;
   signal s1:  std_logic:='0';               end architecture;
   signal f1:  integer;
```

PR 3.20   Program 3.20

The test-bench shown in PR 3.20 can be used for simulation purposes. For this purpose, we can use any simulation platform released by companies, for instance XILINX ISE platform, or XILINX Vivado platform. The test-bench in PR 3.20 can also be written as in PR 3.21 where it is seen that we used the same input port names for the names of the initialization parameters.

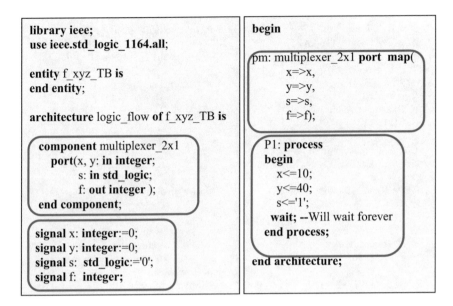

```
library ieee;                          begin
use ieee.std_logic_1164.all;
                                       pm: multiplexer_2x1 port map(
entity f_xyz_TB is                         x=>x,
end entity;                                y=>y,
                                           s=>s,
architecture logic_flow of f_xyz_TB is     f=>f);

   component multiplexer_2x1           P1: process
      port(x, y: in integer;           begin
            s: in std_logic;              x<=10;
            f: out integer );            y<=40;
   end component;                        s<='1';
                                         wait; --Will wait forever
   signal x: integer:=0;               end process;
   signal y: integer:=0;
   signal s:  std_logic:='0';         end architecture;
   signal f:  integer;
```

PR 3.21 Program 3.21

It is possible to create the test-bench using the XILINX ISE platform. However, in this case a template is created, and the data types of the inputs ports of the entity to be tested may not comply to the data types of the component unit created automatically by the ISE platform. In such a case, manual correction should be performed on the created test-bench file. For instance, for the 2×1 integer multiplexer implemented in PR 3.12, the test-bench file automatically created by ISE platform is depicted in PR 3.22.

```
LIBRARY ieee;                        constant <clock>_period : time := 10 ns;
USE ieee.std_logic_1164.ALL;         BEGIN
                                     -- Instantiate the Unit Under Test (UUT)
ENTITY mux_TB IS
END mux_TB;                          uut: multiplexer_2x1 PORT MAP (
                                         x => x,
ARCHITECTURE behavior OF                 y => y,
mux_TB IS                                s => s,
                                         f => f
-- Component Declaration for the Unit     );
Under Test (UUT)                     -- Clock process definitions

  COMPONENT multiplexer_2x1          <clock>_process :process
  PORT(                              begin
    x : IN  std_logic;                        <clock> <= '0';
    y : IN  std_logic;                        wait for <clock>_period/2;
    s : IN  std_logic;                        <clock> <= '1';
    f : OUT  std_logic                        wait for <clock>_period/2;
    );                               end process;
  END COMPONENT;                     -- Stimulus process
                                     stim_proc: process
  --Inputs                          begin
  signal x : std_logic := '0';         -- hold reset state for 100 ns.
  signal y : std_logic := '0';         wait for 100 ns;
  signal s : std_logic := '0';
                                         wait for <clock>_period*10;
  --Outputs
  signal f : std_logic;                 -- insert stimulus here

-- No clocks detected in port list.     wait;
Replace <clock> below with           end process;
  -- appropriate port name
                                     END;
```

PR 3.22 Program 3.22

However, when PR 3.22 is inspected, it is seen that the automatically created test-bench file needs serious modification so that it can be for testing of PR 3.22.

Exercise: Implement the 4 × 1 multiplexer shown in Fig. 3.3 in VHDL and write a test-bench for the testing of the written implementation. Obtain the simulation waveforms using any simulation platform.

Fig. 3.3 4 × 1 multiplexer

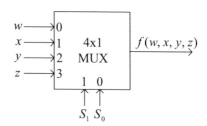

Example 3.3 Implement the 4-bit priority encoder, whose truth table is given in Table 3.2, in VHDL language and write a test-bench for the written implementation.

Table 3.2 Truth table of 4-bit priority encoder

I_3	I_2	I_1	I_0	x	y	z
1	x	x	x	1	1	1
0	1	x	x	1	0	1
0	0	1	x	0	1	1
0	0	0	1	0	0	1
0	0	0	0	x	x	0

Solution 3.3 The priority encoding operation illustrated in Table 3.2 can be implemented as in PR 3.23. For the don't care condition, we use '-' symbol, for instance "1 $\times\times\times$" can be implemented in VHDL as "1 − −−".

```
library ieee;
use ieee.std_logic_1164.all;

entity Four_bit_priority_encoder is
  port( I: in std_logic_vector(3 downto 0);
          F: out std_logic_vector(2 downto 0) );
end entity;

architecture logic_flow of Four_bit_priority_encoder is
begin
with (I) select
  F<="111" when "1---",
      "101" when "01--",
      "011" when "001-",
      "001" when "0001",
      "--0" when  others;
end architecture;
```

PR 3.23 Program 3.23

Test-bench for VHDL program in PR 3.23 can be written as in PR 3.24.

```vhdl
library ieee;
use ieee.std_logic_1164.all;

entity Four_bit_priority_encoder_TB is
end entity;

architecture logic_flow of Four_bit_priority_encoder_TB is

component Four_bit_priority_encoder
  port( I: in std_logic_vector(3 downto 0);
        F: out std_logic_vector(2 downto 0) );
end component;

signal I: std_logic_vector(3 downto 0);
signal F: std_logic_vector(2 downto 0);

begin
pm: Four_bit_priority_encoder PORT MAP(
  I=>I,
  F=>F );

p1: process
begin
        I<="01--";
        wait for 400ns;
        wait;
end process;
end architecture;
```

PR 3.24 Program 3.24

Problems

(1) The truth table of a Boolean function is shown in Table 3.P1. Implement the given truth table in VHDL and write a test-bench for this implementation.

Table 3.P1 The truth table of a Boolean function

x	y	z	$f(x, y, z)$
0	0	0	0
0	0	1	1
0	1	0	0
0	1	1	1
1	0	0	1
1	0	1	0
1	1	0	0
1	1	1	1

(2) The truth table of a Boolean function is shown in Table 3.P2. Implement the
    given truth table in VHDL and write a test-bench for this implementation.

Table 3.P2 The truth table
of a Boolean function

x	y	z	$f(x, y, z)$
0	0	0	1
0	0	1	1
0	1	0	0
0	1	1	0
1	0	0	1
1	0	1	1
1	1	0	0
1	1	1	0

(3) Consider the BCD to Seven Segment Display code converter. Implement the
    BCD to Seven Segment Display code converter in VHDL and write a
    test-bench for this implementation.
(4) Implement the Boolean function $f(w, x, y, z) = y'z' + yz + w'z$ in VHDL, and
    obtain the truth table of the function, and write a test-bench for the VHDL this
    implementation, and use the truth table while deciding on the values of the
    simulation parameters of the test-bench.
(5) The truth table of a Boolean function is shown in Table 3.P3. Implement the
    given truth table in VHDL and write a test-bench for this implementation.

Table 3.P3 The truth table
of a Boolean function
employing four variables

w	x	y	z	$f(x, y, z)$
0	0	0	0	0
0	0	0	1	0
0	0	1	0	0
0	0	1	1	1
0	1	0	0	0
0	1	0	1	1
0	1	1	0	1
0	1	1	1	
1	0	0	0	0
1	0	0	1	1
1	0	1	0	0
1	0	1	1	1
1	1	0	0	1
1	1	0	1	0
1	1	1	0	0
1	1	1	1	1

(6) Obtain the truth table of the logic circuit depicted in Fig. 3.4. Implement the circuit in VHDL and write a test-bench for this implementation.

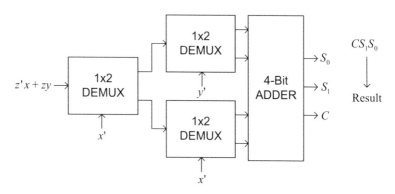

Fig. 3.4 Logic circuit for Problem-5

Chapter 4
User Defined Data Types, Arrays and Attributes

It is vital for a programmer to have a good knowledge of data types and arrays together with the attributes to be able to write qualified programs. In this chapter we will first provide information about defining the data types other than the standard ones available in VHDL libraries, and then the use of arrays together with the attributes will be explained in details with sufficient number of examples.

4.1 User Defined Data Types

In addition to standard data types, such as **integer, std_logic, bit**, etc., we can define our own data types to be used in VHDL programming. The user defined data types are described using the VHDL statement **type**. The syntax of the **type** is as follows:

$$\textbf{type } user\_defined\_data \textbf{ is range } range\_specifications.$$

User defined data types are usually defined in the declarative part of the architecture unit, or it is defined in the packet unit.

Example 4.1 In PR 4.1, we define a new data type and use it to declare a signal object.

© Springer Nature Singapore Pte Ltd. 2019
O. Gazi, *A Tutorial Introduction to VHDL Programming*,
https://doi.org/10.1007/978-981-13-2309-6_4

```
entity type_example is
end entity;

architecture logic_flow of type_example is

-- New data type is defined in the declarative part of the architecture unit
type my_integer  is range  0 to 100;
signal student_mark: my_integer;

begin
  student_mark<=67; -- The assigned number must be between 0 and 100
end architecture;
```

PR 4.1 Program 4.1

4.1.1   Enumerated Types

User defined enumerated data types are usually employed for the implementation of
finite state machines in VHDL language. User defined enumerated data types can be
considered as a set of logic values defined by the user. The syntax of the user
defined enumerated data types is as follows:

$$\textbf{type}\,\text{user\_defined\_data}\,\textbf{is}(\text{value1},\text{value2},\cdots,\text{valueN}).$$

Example 4.2 In PR 4.2, we define a new enumerated data type and use it in the
declaration of a signal object.

PR 4.2 Program 4.2

```
entity enum_type_example is
end entity;

architecture logic_flow of enum_type_example is

type my_logic is ('0', '1', 'Z', 'W');
signal X: my_logic;

begin
  X<='1';
end architecture;
```

Example 4.3 In PR 4.3, we define an enumerated data type called states, and use it in the declaration of a signal object.

PR 4.3 Program 4.3

```
library ieee;
use ieee.std_logic_1164.all;

entity enum_type_example is
 port (x: in std_logic_vector(1 downto 0));
end entity;

architecture logic_flow of enum_type_example is

type states is (state0, state1, state2, state3);
signal currentState, nextState: states;

begin
    currentState<=state0 when x="00" else
                  state1 when x="01" else
                  state2 when x="10" else
                  state3;

    nextState<=currentState;

end architecture;
```

4.2 User Defined Array Data Types

We have already seen the data types, **std_logic_vector**, **bit_vector**, and these vectors are nothing but arrays of **std_logic**, and **bit** data types. We can define the vectors or arrays for the other standard data types or for the user defined data types. Arrays can be defined in two different methods. One of them is the constrained arrays whose declaration includes the explicit declaration of the array size, i.e., the array size, a fixed number, is included in the **type** definition.

4.2.1 Constrained Arrays

The syntax of the user defined constrained array data types is as follows.

type data_array **is array**(definite array range)**of** data_type.

Example 4.4 In PR 4.4, we define an integer vector of length 8, and use it in the body of the architecture part.

```
library ieee;
use ieee.std_logic_1164.all;

entity array_type_example is
end entity;

architecture logic_flow of array_type_example is

type integer_vector is array(7 downto 0) of integer;
signal data: integer_vector;

begin
    data<=(6, 8, 14, 54, 23, 78, 456, 87);
end architecture:
```

PR 4.4   Program 4.4

Example 4.5 In PR 4.5, we define a vector whose elements are **std_logic_vector** of length 8, and use it in the body of the architecture part.

```
library ieee;
use ieee.std_logic_1164.all;

entity array_type_example is
end entity;

architecture logic_flow of array_type_example is

type byte_vector is array(3 downto 0) of std_logic_vector(7 downto 0);
signal data: byte_vector;

begin
    data<=("11011110", "10101011", "00110011", "00111101");
end architecture;
```

PR 4.5   Program 4.5

Accessing to the elements of an array is achieved using the index of the elements. If x is an array, then x(indx) is the element of array at the position indx. Example 4.6 illustrates the concept.

Example 4.6 Accessing to the elements of arrays is illustrated in PR 4.6.

```
library ieee;
use ieee.std_logic_1164.all;

entity array_type_example is
    port(X0, X1, X2, X3, X4, X5, X6, X7: out integer);
end entity;

architecture logic_flow of array_type_example is

type integer_vector is array(7 downto 0) of integer;
signal data1: integer_vector;
signal data2: integer_vector;
signal data3: integer_vector;

begin
    data1<=(6, 8, 14, 54, 23, 78, 456, 87);
    X0<= data1(0); -- X0=87
    X1<= data1(1); -- X1=456
    X2<= data1(2); -- X2=78
    X3<= data1(3); -- X3=23
    X4<= data1(4); -- X4=54
    X5<= data1(5); -- X5=14
    X6<= data1(6); -- X6=8
    X7<= data1(7); -- X7=6
    data2<= data1;
    data3(7)<= data1(7);
    data3(6)<= data1(6);
    data3(5)<= data1(5);
    data3(4)<= 255;
end architecture;
```

PR 4.6 Program 4.6

4.2.2   *Unconstrained Arrays*

In unconstrained array definitions, the size of the array is not a deterministic value, i.e., undefined length. The syntax of the unconstrained array definition is as

$$\textbf{type } \text{data\_array} \textbf{ is array}(\textbf{natural range} <>)\textbf{of} \text{data\_type}$$

where "**natural range <>**" implies that the range limits of the array must fall within the **natural** number range. We can also use **positive**, or **integer** instead of **natural** in array declaration, i.e., we can also have

type data_array **is array**(**positive range** $< >$)**of** data_type.

Example 4.7 In PR 4.7 we define integer vectors of different lengths in the declarative part of the architecture and use them in the body part of the architecture.

```
library ieee;
use ieee.std_logic_1164.all;

entity unconst_array_type_example is
end entity;

architecture logic_flow of unconst_array_type_example is

type int_vector is array (natural range <>) of integer;
signal int_data1: int_vector(3 downto 0);
signal int_data2: int_vector(7 downto 0);

begin
    int_data1<=(2,4,78,1256);
    int_data2<=(45, 56, 34, 7, 7, 67, 45, 34);
end architecture;
```

PR 4.7 Program 4.7

Example 4.8 In PR 4.8 we define two vectors whose elements are logic vectors of length 8.

```
library ieee;
use ieee.std_logic_1164.all;

entity unconst_array_type_example is
end entity;

architecture logic_flow of unconst_array_type_example is
```

```
type byte_array is array (natural range <>) of std_logic_vector(7 downto 0);
signal data_vec1: byte_array(3 downto 0);
signal data_vec1: byte_array(1 downto 0);

begin
    data_vec1<=("00001111","01010101","00000011","01011110");
    data_vec2<=("00001111","01010101");
end architecture;
```

PR 4.8 Program 4.8

4.2.3   Defining PORT Arrays

As we have seen in our previous examples, we can use data types **std_logic_vector**, and **bit_vector** at the port inputs and outputs. If we want to use user defined data vectors at the port inputs and outputs, we need to define these vectors before the **entity** unit of the VHDL program. This is possible if we define user defined vector data types in the **package** unit which comes before the **entity** unit of the VHDL program. The **package** unit can be written in a different file or it can be written in the same file with the **entity** unit. In both cases, we need to add the sentence

use work.package_name.all;

before the **entity** unit.

Example 4.9 Write a VHDL program that inputs an integer vector of length 8 and adds 2 to the every element of the vector and sends it to the output port whose data type is integer vector of length 8.

Solution 4.9 In PR 4.9, we wrote the **package** and **entity** units in the same file. However, after **package** declaration we included the sentence **use work.**my_-data_type.**all** so that we can use the defined data types declared in package unit.

```
package my_data_type is
    type integer_vector is array(natural range <>) of integer;
end package;

use work.my_data_type.all;

entity port_array_type_example is
    port( X: in integer_vector(7 downto 0);
          Y: out integer_vector(7 downto 0));
end entity;

architecture logic_flow of port_array_type_example is

    signal data_vec: integer_vector(7 downto 0);

begin
    data_vec(0)<= X(0)+2;
    data_vec(1)<= X(1)+2;
    data_vec(2)<= X(2)+2;
    data_vec(3)<= X(3)+2;
    data_vec(4)<= X(4)+2;
    data_vec(5)<= X(5)+2;
    data_vec(6)<= X(6)+2;
    data_vec(7)<= X(7)+2;
    Y<=data_vec;
end architecture:
```

PR 4.9 Program 4.9

Example 4.10 Write a VHDL program that inputs a 4 × 4 logic matrix and sends it to an output port.

Solution 4.10 In PR 4.10, we wrote the **package** and **entity** units in the same file. After **package** declaration, we include **use work.**my_data_type**.all** so that we can use the defined data types declared in package unit.

```
library ieee;
use ieee.std_logic_1164.all;

package my_data_type is
   type logic_matrix is array(natural range <>) of std_logic_vector(3 downto 0);
end package;

use work.my_data_type.all;

entity port_array_type_example is
   port( X: in logic_matrix(3 downto 0);
         Y: out logic_matrix(3 downto 0));
end entity;

architecture logic_flow of port_array_type_example is
begin
   Y(0)<= X(0);
   Y(1)<= X(1);
   Y(2)<= X(2);
   Y(3)<= X(3);
   -- Y=X  is also valid
end architecture;
```

PR 4.10   Program 4.10

The body part of the **architecture** unit of PR 4.10 can also be written using the generate statement as in PR 4.11.

PR 4.11   Program 4.11

```
gen: for index in 0 to 3 generate
       Y(index)<= X(index);
     end generate;
```

4.3 Defining 2D Arrays or Matrices

2D arrays can be considered as matrices. We can consider a matrix as a column vector whose elements are nothing but row vectors and vice versa, or we can consider a matrix as table of numbers and the table elements are accessed using the corresponding row and column indexes. The formation of matrix logic is illustrated in Fig. 4.1.

Fig. 4.1 Interpretation of a matrix in different ways

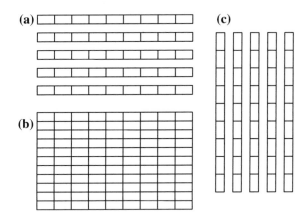

Now, considering a matrix as a column whose elements are row vectors, i.e., row vectors are concatenated in column wise manner as depicted in Fig. 4.1a, we will study its implementation in VHDL language. For such kind of implementation, we will use the **type** statement.

4.3.1 Matrix as Column Wise Concatenated Row Vectors

Example 4.11 Define a matrix data type of size $N \times M$.

Assuming that N and M are static numbers, we can define the logic matrix of size $N \times M$ as in PR 4.12.

```
type row_vector is array(natural range <>) of std_logic;

type logic_matrix is array(natural range <>) of row_vector(M-1 downto 0);

signal data_matrix: logic_matrix(N-1 downto 0);
```

PR 4.12 Program 4.12

Example 4.12 Write a VHDL program that inputs an $N \times M$ logic matrix and sends it to an output port. Consider the matrix as concatenation of row vectors.

Solution 4.12 We cannot get the matrix sizes from input ports. They should be static. We can define them as constant number in the package unit and use them later as in PR 4.13.

```
library ieee;
use ieee.std_logic_1164.all;

package my_data_type is
   constant N: natural:=10;
   constant M: natural:=20;
   type row_vector is array(natural range <>) of std_logic;
   type logic_matrix is array(natural range <>) of row_vector(M-1 downto 0);
end package;

use work.my_data_type.all;

entity matrix_example is
   port( X: in logic_matrix(N-1 downto 0);
         Y: out logic_matrix(N-1 downto 0));
end entity;

architecture logic_flow of matrix_example is
begin
   Y<=X;
end architecture;
```

PR 4.13 Program 4.13

Once we define a matrix as concatenation of row vectors, we can access to the elements or rows of the matrix as in PR 4.14.

```
library ieee;
use ieee.std_logic_1164.all;

entity matrix_example is
end entity;

architecture logic_flow of matrix_example is

    type row_vector is array(natural range <>) of std_logic;
    type logic_matrix is array(natural range <>) of row_vector(2 downto 0);

    signal data_matrix1: logic_matrix(5 downto 0);
    signal data_matrix2: logic_matrix(7 downto 2);
    signal data_matrix3: logic_matrix(10 downto 0);

    signal row_vector1: row_vector (2 downto 0);
    signal row_vector2: row_vector (2 downto 0);

    signal a, b, c: std_logic;

begin

    data_matrix1<=("101", "1ZX", "001", "ZZ1", "010", "111");
    data_matrix2<=(7=>"101", 5=>"111", 3=>('0','1','0'), 2=> "001", others=>"000");

    row_vector1<= data_matrix1(5);
    row_vector2<= data_matrix1(1);

    a<=data_matrix1(5)(1);
    b<=data_matrix1(2)(2);
    c<=data_matrix1(0)(0);

    data_matrix3(5 downto 2)<=data_matrix2(7 downto 4);

    data_matrix3(1)<= row_vector1;
    data_matrix3(4)<= row_vector2;

end architecture;
```

PR 4.14 Program 4.14

4.3.2 Matrix as Table of Numbers

A matrix of size $N \times M$ can be defined using the syntax

type matrixN $\times$ M **is array**(1 **to** N, 1 **to** M)**of data_type**

Example 4.13 Define a matrix of size 3×4. The elements of the matrix are the **std_logic**.

Solution 4.13 The solution is provided in PR 4.15.

```
type matrix3x4  is array(1 to 3, 1 to 4) of std_logic;

signal data_matrix: matrix3x4 :=("0101","0011","1110");
```

PR 4.15 Program 4.15

Example 4.14 In PR 4.16, we define a constant matrix of size 3×4 and read its elements.

```
library ieee;
use ieee.std_logic_1164.all;

entity matrix_example is
  port(column_indx: in natural range 1 to 4;
       row_indx: in natural range 1 to 3;
       column_vector: out std_logic_vector(1 to 3);
       row_vector: out std_logic_vector(1 to 4);
       elem: out std_logic);
end entity;

architecture logic_flow of matrix_example is

    type matrix3x4  is array(1 to 3, 1 to 4) of std_logic;
    constant data_matrix: matrix3x4 :=(( '0', '1', '0', '1'),
                                        ('0', '0', '1', '1'),
                                        ('1', '1', '1', '0'));
begin
  row_vector<= data_matrix(row_indx, 1 to 4);
  column_vector<= data_matrix(1 to 3, column_indx);
  elem<= data_matrix(row_indx, column_indx);
end architecture;
```

PR 4.16 Program 4.16

Example 4.15 We can define a matrix of size 3 × 4 using unconstrained matrix declaration as in PR 4.17. The elements of the matrix are the **std_logic**.

Solution 4.15 The solution is provided in PR 4.17.

---

type matrix **is array (natural** <>, **natural range** <>) **of std_logic**;

signal data_matrix3x4: **matrix(2 downto** 0, 3 **downto** 0);

data_matrix3x4<= (('0', '1', '0', '1'),
              ('0', '0', '1', '1'),
              ('1', '1', '1', '0'));

PR 4.17 Program 4.17

4.3.3 3D Arrays

Vector of 2D Arrays, i.e., array of 2D arrays, can be considered as 3D arrays. Following the same logic, we can also define 4D, 5D, ... arrays. Let's illustrate the concept with an example.

3D arrays can be defined with two different approaches. In the first approach, we define the 3D array via vector of matrices. In the second approach, we define 3D arrays as 3 dimensional table of integers.

Example 4.16 Declaration of 2D, 3D, and 4D arrays can be done as in PR 4.18.

type integer_vector **is array (natural range** <>) **of integer**;

type integer_matrix **is array (natural range** <>) **of integer_vector**(2 **downto** 0);

type twoD_array **is array (natural range** <>) **of integer_vector**(2 **downto** 0);

type threeD_array **is array (natural range** <>) **of twoD_array** (2 **downto** 0);

type threeD_array2ndM **is array (natural range** <>, **natural range** <>,
                                              natural range <>) **of integer**;

type fourD_array **is array (natural range** <>) **of threeD_array** (3 **downto** 0);

signal data_vector: **integer_vector**(2 **downto** 0) :=(5, 6, 23);

signal data_2D: **integer_matrix**(2 **downto** 0) :=((5, 6, 23),(1, 4, 6),(9, 23, 67));

signal data_3D: **threeD_array** (1 **downto** 0) :=(((5, 6, 23),(1, 4, 6),(9, 23, 67)),
                                                           ((5, 6, 23),(1, 4, 6),(9, 23, 67)));

signal data1_3D: **threeD_array** (2 **downto** 0, 2 **downto** 0, 2 **downto** 0) :=
                                                                  ((5, 6, 3),
                                                                    (1, 4, 6),
                                                                    (9, 2, 7));

PR 4.18 Program 4.18

Example 4.17 In PR 4.19, **type** declarations for complex number arrays are given. The real and imaginary parts of the complex numbers are integers.

type real_im **is array**(1 **downto** 0) **of integer**;
type complex_num_array **is array (natural range** <>) **of** real_im;
signal complex_vector: complex_num_array(63 **downto** 0);

PR 4.19 Program 4.19

4.4 Subtypes

A subtype declaration does not introduce a new type. Although operations are not allowed between different data types, they are allowed between a type and its subtype. Subtype data declarations are usually done in the declarative part of the architecture or in the package unit. The syntax of the subtype declaration is as:

subtype subtype_name **is** base_type **range** range_description;

```
type my_integer  is range  0 to 1000;
subtype small_integer is my_integer range 0 to 20;
signal data1: small_integer;
-----------------------------------------------------------------------
type digits is (0,1,2,3,4,5,6,7,8,9) ;
subtype middle_digits is digits range 3 to 6;

signal data2: middle_digits:=4;
signal data3: digits;
data3<=data2;
-----------------------------------------------------------------------
subtype nibble is std_logic_vector(3 downto 0);
type memory is array(2 downto 0, 2 downto 0) of nibble;
signal data4 : memory;
data4<=( ("1010", "1111", "0011"),
         ("1010", "1111", "0011"),
         ("1010", "1111", "0011") );
```

PR 4.20 Program 4.20

The type of a subtype is the same type as its base type. When objects of a subtype and its base type are assigned to each other, no type conversion is needed.

Example 4.18 In PR 4.20, we provide some examples for **subtype** declarations.

4.4.1 Type Conversion

A binary string may represent, a **std_logic_vector**, an **unsigned** bit vector, a **signed** bit vector, an **integer**, a **natural**, or a **positive** data. It is possible to convert a binary string, or bit vector, from a data type to the other via casting operation. The following example illustrates the concept clearly.

Example 4.19 In PR 4.21, type conversion operation is illustrated. In PR 4.21 we included **numeric_std** package at the top of the program. This packet is needed for the conversions from **std_logic_vector** to **unsigned** from **std_logic_vector** to **signed** data types.

```
library ieee;
use ieee.std_logic_1164.all;
use ieee.numeric_std.all;

entity type_conversion is
  port( logic_vec1: in std_logic_vector(4 downto 0);
          unsigned_vec: out unsigned(4 downto 0);
          signed_vec: out signed(4 downto 0);
  end entity;

architecture logic_flow of type_conversion is
begin
    unsigned_vec<=unsigned(logic_vec1);
    signed_vec<=signed(logic_vec1);
end architecture;
```

PR 4.21 Program 4.21

Example 4.20 If a, b and c are of **signed** data type and d is of **std_logic_vector** data type, then for the conversion

$$d <= \textbf{std\_logic\_vector}(a * b)$$

we need the **numeric_std** package in our program. On the other hand, if

$$c <= a * b;$$

then for the conversion

$$d <= \textbf{std\_logic\_vector}(c);$$

we need to include **std_logic_arith** package in our program.

There are also type conversion functions available. Some of these functions can be outlined as follows:

to_unsigned$(\cdot)$, **to_signed**$(\cdot)$, **to_integer**$(\cdot)$, **to_bitvector**$(\cdot)$, **to_stdlogicvector**$(\cdot)$

4.5   Attributes of Data Types

Attributes can be considered for scalar, array, and signal objects separately.

4.5.1   *Attributes for Scalar Data Types*

We can define a scalar data type using

type T **is range** range_specifications.

such as

type student_mark **is range** 0 **to** 100;

or we can already use a pre-defined data type T such as **natural**. For a scalar data type T, we have the attributes in Table 4.1.

Table 4.1 Attributes for scalar data types

Attribute	Result
T'left	is the leftmost value of data type T
T'right	is the rightmost value of data type T
T'high	is the largest value of data type T
T'low	is the smallest value of data type T
T'pos(x)	is the index of value **x** in the data type T
T'val(x)	is the value whose index in the data type T is **x**
T'value(x)	is the value of data type T converted from string **x**
T'leftof(x)	is the value one position to the left of **x**
T'rightof(x)	is the value one position to the right of **x**
T'succ(x)	is the value at position one greater than position of **x** in T
T'pred(x)	is the value at position one less than position of **x** in T
T'ascending	is true of the range of T is ascending, false otherwise
T'base	is the base type of the type T
T'image(x)	is a string representation of **x** of data type T

For ascending ranges, we have

$$T'\textbf{succ}(x) = T'\textbf{rightof}(x)$$
$$T'\textbf{pred}(x) = T'\textbf{leftof}(x)$$

and for descending ranges, we have

$$T'\textbf{succ}(x) = T'\textbf{leftof}(x)$$
$$T'\textbf{pred}(x) = T'\textbf{rightof}(x)$$

Example 4.21 In PR 4.22, we define a new data type and use it to illustrate the attributes.

type states **is** (st0, st1, st2, st3); **signal** currSt: **states**; **signal** flag: **boolean**; **signal** index: **integer**; currSt<= states'**left** ; -- currSt=st0 currSt<= states'**right**; -- currSt=st3 currSt<= states'**low**; -- currSt=st0 currSt<= states'**high** ;-- currSt=st3	flag<= states'**ascending** ; -- flag= true; states'**image**(st1); -- result is "st1" index<= states'**pos**(st2); -- index= 2 currSt<= states'**succ**(st2); -- currSt=st3 currSt<= states'**pred**(st3); -- currSt=st2 currSt<= states'**leftof**(st3); -- currSt=st2 currSt<= states'**rightof**(st2); -- currSt=st3

PR 4.22 Program 4.22

Example-4.22 In PR 4.23, we define a new data type and use it to illustrate the attributes.

type my_integer **is range** 31 **downto** -6; **signal** x_int : **my_integer**; **signal** flag: **boolean**; **signal** index: **integer**; x_int <= my_integer'**left**; -- x_int=31 x_int <= my_integer'**right**; -- x_int=-6 x_int <= my_integer'**low**; -- x_int=-6 x_int <= my_integer'**high**;-- x_int=31	flag<= my_integer'**ascending**; -- false x_int <= my_integer'**value**("18"); -- x_int=18 x_int <= my_integer'**val**(3); -- x_int=28 x_int <= my_integer'**succ**(14); -- x_int=15 x_int <= my_integer'**pred**(14) ; -- x_int=13 x_int <= my_integer'**leftof**(16); -- x_int=17 x_int <= my_integer'**rightof**(10); -- x_int=9

PR 4.23 Program 4.23

Note that in PR 4.22 and PR 4.23 we made multiple assignments to the same signal object for illustrative purposes, in fact, only a signal assignment is allowed to a signal object in the body of the architecture unit.

4.5.2 Attributes for Array Data Types

Let A be an object whose data type is array. Then, for this object we have the attributes listed in Table 4.2.

Table 4.2 Attributes for array data types

Attribute	Result
A'left	leftmost index of array range
A'right	rightmost index of array range
A'low	smallest index value of the array range
A'high	largest index value of the array range
A'range	index range of the array
A'reverse_range	reverse of the array range
A'length	the length of the array range
A'ascending	true if index range is ascending, otherwise it is false

Example 4.23 In PR 4.24, we define a new array data type and use it to illustrate the attributes.

```
type integer_vector is array(natural range <>) of integer;
signal int_vec : integer_vector(7 downto 0);

signal data: std_logic_vector(int_vec'range);          -- 7 downto 0
signal temp : std_logic_vector( (int_vec'length–1) downto 0); -- 7 downto 0

signal rev_data : std_logic_vector (int_vec'reverse_range);   -- 0 to 7

signal num : integer range 0 to 7;
signal flag : boolean;

num<=int_vec'left;  -- num =7
num<=int_vec'low;  -- num =0
num<=int_vec'right; -- num =0
num<=int_vec'high; -- num =7

flag<= int_vec'ascending; -- flag =false
flag<= rev_data'ascending; -- flag=true

my_sig<= x"cc";
rev_sig<=my_sig; -- x"33"
```

PR 4.24 Program 4.24

A multi-dimensional array A have the attributes listed in Table 4.3.

Table 4.3 Attributes for multi-dimensional arrays

Attribute	Result
A'left(n)	leftmost index of array range for the nth dimension
A'right(n)	rightmost index of array range for the nth dimension
A'low(n)	smallest index value of the array range for the nth dimension
A'high(n)	largest index value of the array range for the nth dimension
A'range(n)	index range of the array for the nth dimension
A'reverse_range(n)	reverse of array range for the nth dimension
A'length(n)	the length of array range for the nth dimension
A'ascending(n)	for the nth dimension, true for ascending array, otherwise it is false

Example 4.24 In PR 4.25, we define two dimensional array data type and use it to illustrate the attributes.

```
type int_matrix is array(0 to 3, 7 downto 0)        matrix'low(1); -- is 0
of integer;                                         matrix'low(2); -- is 0
signal matrix: int_matrix;
                                                    matrix'range(1); -- is 0 to 3
matrix'left(1); -- is 0                              matrix'range(2); -- is 7 downto 0
matrix'left(2); -- is 7
                                                    matrix'reverse_range(1);-- is 3 downto 0
matrix'right(1); -- is 3                             matrix'reverse_range(2); -- is 0 to 7
matrix'right(2); -- is 0
                                                    matrix'length(1); -- is 4
matrix'high(1); -- is 3                              matrix'length(2); -- is 8
matrix'high(2); -- is 7
```

PR 4.25 Program 4.25

Example 4.25 In PR 4.26, part of the previous example is available in a complete VHDL program.

```
entity array_attributes is
end entity;

architecture logic_flow of array_attributes is
    type int_matrix is array(0 to 3, 7 downto 0) of integer;
    signal matrix: int_matrix;
    signal n1,n2, n3, n4, n5, n6: integer;

begin
        n1<=matrix'left(1); -- is 0
        n2<=matrix'left(2); -- is 7

        n3<=matrix'right(1); -- is 3
        n4<=matrix'right(2); -- is 0

        n5<=matrix'high(1); -- is 3
        n6<=matrix'high(2); -- is 7
end architecture:
```

PR 4.26 Program 4.26

4.5.3 Attributes for Signal Objects

If S is a signal object, then we can define the attributes in Table 4.4 for the given signal object.

Table 4.4 Attributes for signal objects

Attribute	Result
S'active	true if a transaction is scheduled for S, false otherwise
S'last_event	is the amount of time passed since the last event on signal S
S'last_active	is the amount of time passed since signal S was last active
S'last_value	is the previous value of signal S before the last event
S'event	is true if an event has occurred on S in the current simulation cycle
S'transaction	is a bit that toggles as S becomes active in the current simulation cycle
S'quiet	is true if no event occurs on S during the current simulation cycle
S'stable	is true if no event has occurred on S during the current simulation cycle
S'delayed(T)	is the value of S at time t-T, i.e., value of the S delayed by T
S'stable(T)	is true if no event has occurred on S during the last T units of time
S'quiet(T)	is true if there has been no transaction on S during the last T units of time

Example 4.26 In PR 4.27, different VHDL statements for the rising edge detection of a clock signal are provided.

PR 4.27 Program 4.27

```
signal clk: std_logic;

when(clk'event and clk='1')..........;
when(not clk'stable and clk='1') ..........;
when(clk'event and clk'last_value='0') ..........;
when(not clk'stable and clk'last_value='0') ..........;
```

From the VHDL statements in PR 4.27 for the rising edge detection of a clock signal, the most commonly used one is

$$\text{clk}'\textbf{event and } \text{clk} = \,'1'$$

Example 4.27 In PR 4.28, different VHDL statements for the falling edge detection of a clock signal are provided.

PR 4.28 Program 4.28

```
signal clk: std_logic;

when(clk'event and clk='0')..........;
when(not clk'stable and clk='0') ..........;
when(clk'event and clk'last_value='1') ..........;
when(not clk'stable and clk'last_value='1') ..........;
```

Problems

(1) Define a data type for complex numbers.
(2) Define an array for a complex number vector.
(3) What is the difference between a constrained array and an unconstrained array.
(4) Define an unconstrained 3D array of complex numbers.
(5) Define a matrix for complex numbers.
(6) Find the mistake in the following declaration.

$$\textbf{type } \text{int\_vector } \textbf{is array}(\textbf{range} < >)\textbf{of integer};$$

(7) Is there any mistake in the following declaration

$$\textbf{type } \text{int\_vector } \textbf{is array}(-3 \textbf{ to } 7)\textbf{of integer};$$

(8) Consider the VHDL program in PR 4.P1, trace the program and find the values of row_vector1, row_vector2, a, b, c, data_matrix3(10 **downto** 7).

```vhdl
library ieee;
use ieee.std_logic_1164.all;

entity matrix_example is
end entity;

architecture logic_flow of matrix_example is

    type row_vector is array(natural range <>) of integer;
    type logic_matrix is array(natural range <>) of row_vector(2 downto 0);

    signal data_matrix1: logic_matrix(4 downto 0);
    signal data_matrix2: logic_matrix(8 downto 1);
    signal data_matrix3: logic_matrix(13 downto 0);

    signal row_vector1: row_vector (2 downto 0);
    signal row_vector2: row_vector (2 downto 0);

    signal a, b, c: integer;

begin
    data_matrix1<=((3,5,6),( 7,5,8), (3,75,76), (9,67,23), (82,67,10) );
    data_matrix2<=(7=>( 3,5,6), 5=( 8,2,5),3=>( 4,4,4), others=>(0,0,0));

    row_vector1<= data_matrix1(3);
    row_vector2<= data_matrix1(2);

    a<=data_matrix1(3)(1);
    b<=data_matrix1(2)(2);
    c<=data_matrix1(1)(0);

    data_matrix3(10 downto 7)<=data_matrix2(7 downto 4);

    data_matrix3(1)<= row_vector1;
    data_matrix3(2)<= row_vector2;

end architecture;
```

PR 4.P1   Program 4P1

Chapter 5
Sequential Circuit Implementation in VHDL

In this chapter we will explain the implementation of sequential logic circuits in VHDL. Implementation of clocked or sequential logic circuits differ from the implementation of combinational logic circuits in the type of VHDL coding. The sequential logic circuits as its name implies works in a sequential manner, and the sequence of operations are controlled by a periodic signal called **clock**. The VHDL codes written for the implementation of the sequential logic circuits works in a sequential manner. On the other hand, the VHDL codes written for the implementation of the combinational logic circuits work in a parallel manner, i.e., work in a concurrent manner.

5.1 Sequential Circuits in VHDL

Digital logic circuits can be divided into two main categories. The first one is the combinational logic circuits, the second one is the synchronous sequential logic circuits, i.e., clocked circuits.

In concurrent coding, we can use the VHDL statements **when, select..with, generate**. On the other hand, in sequential coding, the statements **if**, **wait**, **loop**, and **case** are employed.

The implementation of the combinational logic circuits is done in the architecture unit of the VHDL program, and combinational circuits are implemented using concurrent VHDL codes. On the other hand, sequential logic circuits are implemented using sequential program units. These sequential units are **process, function**, and **procedure**. The **process** unit is usually considered to be used in the architecture part of the VHDL module, whereas, the **function**, and **procedure** units are usually employed inside the **packet** unit.

© Springer Nature Singapore Pte Ltd. 2019
O. Gazi, *A Tutorial Introduction to VHDL Programming*,
https://doi.org/10.1007/978-981-13-2309-6_5

5.1.1 Process

The syntax of the **process** unit is as illustrated in PR 5.1.

PR 5.1 Program 5.1

> **process**(sensitivity list)
>   declaration part of the process
> **begin**
>   body part of the process
> **end process**;

A **process** is used to implement a sequential logic circuit. Whenever a parameter in the sensitivity list of the process changes, the **process** is executed. The **variable** object if needed can be used in the declaration and body part of the process. Declaration of the signal objects are not allowed inside processes.

As mentioned previously **if**, **wait**, **for**, **loop**, and **case** statements are used inside the sequential program units.

5.1.2 IF Statement

The **if** statement is used to check the conditions. The syntax of the **if** statement is as follows:

if conditions **then**

Statements;

elsif conditions **then**

Statements;

⋮

else

Statements;

end if;

Let's see the implementation of some of the sequential logic circuits in VHDL language.

5.2   Example Implementations for Sequential Logic Units

In this section, we will provide some examples for the implementation of some well-known clocked logic units.

5.2.1   D-Type Flip Flop

Example 5.1 The symbolic representation of the positive edge triggered D-type flip flop is shown in Fig. 5.1.

Fig. 5.1 The symbolic representation of the positive edge triggered D-type flip flop

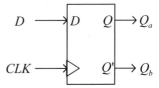

The characteristic table of the D-type flip flop is depicted in Table 5.1.

Table 5.1 The characteristic table of the D-type flip flop

CLK	D	Q_a	Q_b
↑	0	0	1
↑	1	1	0
0	×	Q	Q'
1	×	Q	Q'
↑	×	Q	Q'

Implement the positive edge triggered D-type flip flop in VHDL language.

Solution 5.1 The implementation of the positive edge triggered D-type flip flop is given in PR 5.2.

```
library ieee;
use ieee.std_logic_1164.all;

entity D_FlipFlop is
    port(d: in std_logic;
          clk: in std_logic;
          qa: out std_logic;
          qb: out std_logic);
end entity;

architecture logic_flow of D_FlipFlop is
begin
    process(clk)
    begin
        if (clk'event and clk='1') then
            qa<=d;
            qb<=not d;
        end if;
    end process;
end architecture;
```

PR 5.2 Program 5.2

The triggering condition

$$\textbf{if } (\text{clk}'\textbf{event and } \text{clk} = \text{`1'}) \textbf{ then}$$

of the D-type flip-flop can be replaced by

$$\textbf{if } (\text{rising\_edge}(\text{clk})) \textbf{ then}$$

In this case, PR 5.2 happens to be as in PR 5.3.

PR 5.3 Program 5.3

```
library ieee;
use ieee.std_logic_1164.all;

entity D_FlipFlop is
    port(d, clk: in std_logic;
            qa, qb: out std_logic );
end entity;

architecture logic_flow of D_FlipFlop is
begin
    process(clk)
    begin
        if (rising_edge(clk)) then
            qa<=d;
            qb<=not d;
        end if;
    end process;
end architecture:
```

Example 5.2 If a reset control is added to the D-type flip flop as shown in Fig. 5.2, then the implementation of the D-type flip flop with synchronous reset can be done as in PR 5.4 and with asynchronous reset it can be implemented as in PR 5.5.

Fig. 5.2 D-type flip flop with reset input

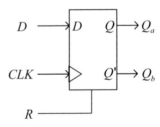

PR 5.4 Program 5.4

```
library ieee;
use ieee.std_logic_1164.all;

entity D_FlipFlop is
    port(d, clk, reset: in std_logic;
         qa, qb: out std_logic );
end entity·
architecture logic_flow of  D_FlipFlop is
begin
    process(clk, reset)
    begin
        if (rising_edge(clk)) then
            if (reset='1') then
                qa <= '0';   qb <= '1';
            else
                qa<=d;   qb<=not d;
            end if;
        end if;
    end process;
end architecture:
```

PR 5.5 Program 5.5

```
library ieee;
use ieee.std_logic_1164.all;

entity D_FlipFlop is
    port(d, clk, reset: in std_logic;
         qa, qb: out std_logic );
end entity;

architecture logic_flow of  D_FlipFlop is
begin
    process(clk, reset)
    begin
            if (reset='1') then
                qa <= '0';   qb <= '1';
            elsif (rising_edge(clk)) then
                qa<=d;   qb<=not d;
            end if;
    end process;
end architecture;
```

Example 5.3 If the D-type flip flop includes both reset and set controls as shown in Fig. 5.3, then the implementation of the D-type flip flop happens to be as in PR 5.6.

Fig. 5.3 D-type flip flop with reset and set inputs

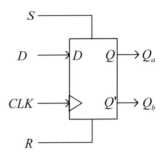

PR 5.6 Program 5.6

```
library ieee;
use ieee.std_logic_1164.all;

entity D_FlipFlop is
    port(d, clk, reset, set: in std_logic;
         qa, qb: out std_logic );
end entity;

architecture logic_flow of D_FlipFlop is
begin
    process(clk, reset, set)
    begin
        if (reset='1') then
            qa <= '0';   qb <= '1';
        elsif(set='1')
            qa <= '1';   qb <= '0';
        elsif (rising_edge(clk)) then
            qa<=d;   qb<=not d;
        end if;
    end process;
end architecture;
```

We can, although not very necessary, use the **if** statement for the implementation of the combinational circuits in VHDL language. Let's illustrate this with an example.

5.2.2 Multiplexer

Example 5.4 Implement the 4×1 multiplexer shown in Fig. 5.4 using the **if-then-else** statement.

Fig. 5.4 4×1 multiplexer symbolic representation

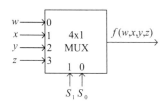

Solution 5.4 The implementation of 4×1 multiplexer shown in Fig. 5.4 is depicted in PR 5.7 with a process unit.

PR 5.7 Program 5.7

```
library ieee;
use ieee.std_logic_1164.all;

entity multiplexer_4x1 is
  port( w, x, y, z, s1, s0: in std_logic;
        f: out std_logic );
end entity;

architecture logic_flow of multiplexer_4x1 is
begin
  process(w, x, y, z, s1, s0)
    begin
      if (s1='0' and s0='0') then f<=w;
      elsif (s1='0' and s0='1') then f<=x;
      elsif (s1='1' and s0='0') then f<=y;
      else f<=z;
      end if;
    end process;
end architecture;
```

5.2.3 JK *and* T *Flip-Flops*

Example 5.5 The black box representation of positive and negative edge triggered *JK* and *T* flip-flops are depicted in Fig. 5.5, and the excitation tables of *JK* and *T* flip-flops are shown in Table 5.2. Implement the *JK* and *T* flip-flops in VHDL.

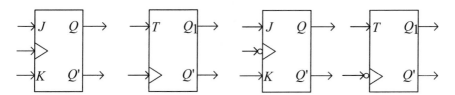

Fig. 5.5 *JK* and *T* flip-flops

Table 5.2 Excitation lists of *JK* and *T* flip-flops

J	K	$Q(t+1)$		T	$Q(t+1)$
0	0	$Q(t)$		0	$Q(t)$
0	1	0		1	$Q'(t)$
1	0	1			
1	1	$Q'(t)$			

Solution 5.5 The implementation of positive edge triggered *JK* flip flop is provided in PR 5.8.

PR 5.8 Program 5.8

```
library ieee;
use ieee.std_logic_1164.all;

entity JK_FlipFlop is
    port(j, k, reset, clk: in std_logic;
            qa, qb: out std_logic );
end entity;

architecture logic_flow of  JK_FlipFlop is
  signal q_temp: std_logic;
begin
    process(clk, reset)
    begin
            if (reset='1') then
                q_temp<='0';
            elsif (rising_edge(clk)) then
                if(j='0' and k='0') then
                    q_temp <= q_temp ;
                elsif(j='0' and k='1') then
                    q_temp <='0';
                elsif(j='1' and k='0') then
                    q_temp <='1';
                else
                    q_temp <= not q_temp;
                end if;
            end if;
        end process;
        qa<= q_temp; qb<=not q_temp;
end architecture;
```

The implementation of positive edge triggered *T* flip flop is provided in PR 5.9.

PR 5.9 Program 5.9

```
library ieee;
use ieee.std_logic_1164.all;

entity T_FlipFlop is
    port(t, reset, clk: in std_logic;
            qa, qb: out std_logic );
end entity;
architecture logic_flow of T_FlipFlop is
    signal q_temp: std_logic;
begin
    process(clk, reset)
    begin
            if (reset='1') then
                q_temp<='0';
            elsif (rising_edge(clk)) then
            if(t='0') then
                q_temp<=q_temp;
            else
                q_temp<= not q_temp;
            end if;
            end if;
        end process;
        qa<= q_temp; qb<=not q_temp;
end architecture;
```

The implementation of negative edge triggered *JK* and *T* flip-flops is very similar to PR 5.9. The only exception is that instead of "rising_edge(clk)", we should use "falling_edge(clk)" in the "if" statement.

5.2.4 Counter

An *N*-bit counter, whose symbolic representation is depicted in Fig. 5.6, is a digital circuit that increments the integer at its output at every clock pulse.

Fig. 5.6 Symbolic representation of *N*-bit counter

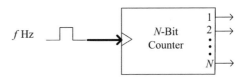

Example 5.6 Implement a 4-bit counter in VHDL language.

Solution 5.6 A 4-bit counter can be implemented in a number of ways as in PR 5.10, 5.11, and 5.12.

PR 5.10 Program 5.10

```vhdl
library ieee;
use ieee.std_logic_1164.all;

entity up_counter is
   port ( clk: in std_logic;
          reset: in std_logic;
          number: out natural range 0 to 15 );
end entity;

architecture logic_flow of up_counter is
   signal temp_num: natural range 0 to 15;
begin
 process(clk,reset)
 begin
    if (reset='1') then
       temp_num <= 0;
    elsif (rising_edge(clk)) then
       temp_num <= temp_num + 1;
    end if;
  end process;
  number <= temp_num;
end architecture;
```

PR 5.11 Program 5.11

```vhdl
library ieee;
use ieee.std_logic_1164.all;

entity up_counter is
   port ( clk, reset: in std_logic;
          number: out natural range 0 to 15 );
end entity;

architecture logic_flow of up_counter is
begin
 process(clk,reset)
    variable temp_num: natural range 0 to 15;
 begin
    if (reset='1') then
       temp_num := 0;
    elsif (rising_edge(clk)) then
       temp_num := temp_num + 1;
    end if;
    number<= temp_num;
  end process;
end architecture;
```

```
library ieee;
use ieee.std_logic_1164.all;
use ieee.std_logic_unsigned.all;

entity up_counter is
    port ( clk, reset: in std_logic;
           number: out std_logic_vector(3 downto 0) );
end entity;
architecture logic_flow of up_counter is
    signal    temp_num:      std_logic_vector(3 downto 0):="0000";

begin

  process(clk,reset)
  begin
    if (reset='1') then
        temp_num <= "0000";
    elsif (rising_edge(clk)) then
        temp_num <= temp_num + 1;
    end if;
  end process;
number <= temp_num;
end Architecture;
```

PR 5.12   Program 5.12

If we use **std_logic_vector** for counter data type as in PR 5.12, then we need to initialize it at the declarative part of the architecture to see the simulation results. Otherwise, we cannot observe the counter increment. Since, the default value of any data type is its leftmost value. For **std_logic**, the leftmost value is 'U'.

5.2.5   *Clock Divider (Frequency Divider)*

Clock generation at the desired frequency is a very important concept for the design of sequential logic units. For this reason, we advise the reader study and understand very well the topics explained in this section.

We can use the counters to generate different square pulse trains with different frequencies. The output pins of an l-bit counter can be used as a frequency divider circuit. If the clock frequency of the counter is f Hz, then at the most significant bit, i.e., $(l-1)^{th}$ bit, of the counter, we have a pulse train of frequency $f/2^l$. This situation is illustrated in Fig. 5.7.

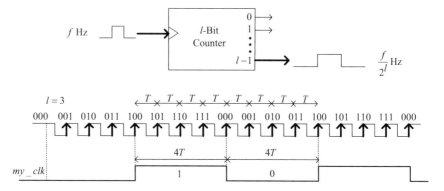

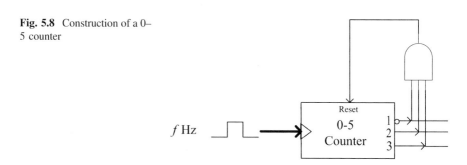

Fig. 5.7 Frequency division using an *l*-bit counter

Using l bits we can generate the integers in the range 0 to 2^{l-1}. For instance when $l = 3$, we can generate the integers $0, 1, 2, 3, \ldots, 7$. To construct a 0 to $K - 1$ counter such that $K < 2^{l-1}$, the counter device needs a reset when the binary equivalent of integer K is available at the l-bit counter output. For instance, for $l = 3$, a 0–5 counter can be constructed as in Fig. 5.8.

Fig. 5.8 Construction of a 0–5 counter

A frequency divider can be constructed using a 0 to $K - 1$ counter. For instance using a 0–5 counter we can construct a frequency divider circuit as in Fig. 5.9 where the synchronous reset operation is performed, i.e., reset operation is performed when counter value reaches 6. This means for the when counter value reaches to 5, we wait for the next clock pulse for the reset operation.

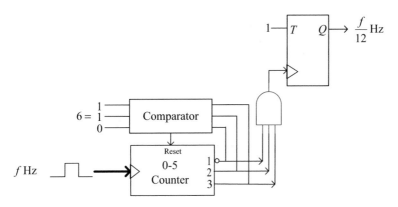

Fig. 5.9 Frequency division operation using a 0–5 counter

The timing waveform of the circuit in Fig. 5.9 is depicted in Fig. 5.10 where it is seen that using a 0–5 counter we can generate a frequency of

$$\frac{f}{12}\,\text{Hz}$$

from an f Hz pulse train. We can generalize this result as: using a 0 to $K-1$ counter we can generate a

$$\frac{f}{2K}\,\text{Hz}$$

clock source from an f Hz pulse train.

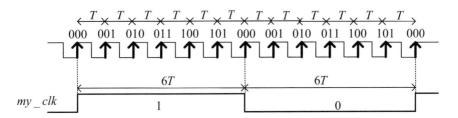

Fig. 5.10 Frequency divider timing waveform using a 0–5 counter

In fact, if Figs. 5.9 and 5.10 are inspected, we see that our counter counts up to the 6, but the value 6 is never seen at the counter output. The reason for this situation is that when the value 6 is reached the counter is reset. For this reason, the value of 6 is never seen at the counter output. Considering this, we call this counter as 0–5 counter. Since only those values from 0 to 5 are seen at the counter output.

A less complex circuit for a general frequency divider involving less logic units is depicted in Fig. 5.11. The timing waveform of this circuit for a 0–5 counter is the same as the waveform of Fig. 5.10.

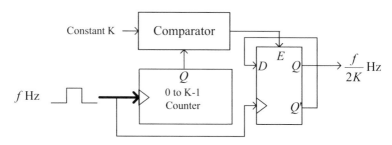

Fig. 5.11 Frequency divider circuit

A frequency divider with a 0 to $K - 1$ counter equals to a frequency divider with a 1 to K counter. Hence, we can state that using a frequency divider with either a 0 to $K - 1$ counter or with a 1 to K counter it is possible to generate a clock source of $\frac{f}{2K}$ Hz from an f Hz supply.

Example 5.7 We have a frequency divider. When the counter value is reached to 25 a reset operation is performed. Find the frequency of the divider output.

Solution 5.7 Our counter repeats the count sequence $0, 1, 2, \ldots, 24$. From this sequence we see that

$$K - 1 = 24 \rightarrow K = 25.$$

Then, out frequency divider output is

$$\frac{f}{2K} \rightarrow \frac{f}{50}.$$

Frequency Divider in VHDL

Frequency division operation in VHDL shows some differences considering the use of signal and variable objects for counter index. The value assignment to a signal object is not immediate, and for this reason, a signal object used for a counter index creates more delay considering the same count sequence compared to a variable object used for counter index. When we mention a 1 to K counter, we mean the at the counter output the sequence $1, 2, \ldots, K$ appears repeatedly.

Let f be the frequency of the FPGA device. If we use a signal object for the counter index, then a 0 to $K - 2$ counter can generate a clock of

$$\frac{f}{2K}.$$

or a 1 to $K - 1$ counter can generate a clock of

$$\frac{f}{2K}.$$

If we use a variable object for counter index, then a 0 to $K - 1$ counter can generate a clock of

$$\frac{f}{2K}.$$

or a 1 to K counter can generate a clock of

$$\frac{f}{2K}.$$

Example 5.8 Design a frequency divider that divides the clock frequency of the FPGA device by 10.

Solution 5.8 We can write this program using many different approaches. If we use a signal object for the counter, and if our counter repeats the sequence 1,2,3,4, then by toggling a logic value when counter value reaches to 5 and initializing the counter value to 1 again, we can obtain the desired frequency.

The required frequency divider using a signal object is implemented in VHDL in PR 5.13.

```
library ieee;
use ieee.std_logic_1164.all;

entity frequency_divider_by_10 is
   port ( clk_in, reset: in std_logic;
          clk_out: out std_logic);
end entity;

architecture logic_flow of frequency_divider_by_10 is
   signal count: positive range 1 to 5;
   signal temp_clk_out: std_logic:='0';
begin
  process(clk_in,reset)
  begin
    if(reset='1') then
       temp_clk_out <='0';
       count<=1;
    elsif (rising_edge(clk_in)) then
       count <= count + 1;
       if(count=5) then
          temp_clk_out<=not temp_clk_out;
          count<=1;
       end if;
    end if;
  end process;
 clk_out<= temp_clk_out;
end architecture;
```

PR 5.13 Program 5.13

On the other hand, the same goal can be achieved using a variable object employed for counter index. In this case, our counter repeats the sequence 1,2,3,4,5, and when the counter value reaches to 6, we toggle a logic value and initialize the counter value to 1 again. The frequency divider with variable object is given in PR 5.14.

```
library ieee;
use ieee.std_logic_1164.all;

entity frequency_divider_by_10 is
    port ( clk_in, reset: in std_logic;
            clk_out: out std_logic);
end entity;

architecture logic_flow of frequency_divider_by_10 is
    signal temp_clk_out: std_logic:='0';
begin
  process(clk_in,reset)
    variable count: positive range 1 to 6;
  begin
    if(reset='1') then
       temp_clk_out <='0';
       count:=1;
    elsif (rising_edge(clk_in)) then
       count := count + 1;
       if(count=6) then
          temp_clk_out<=not temp_clk_out;
          count:=1;
       end if;
    end if;
  end process;
  clk_out<= temp_clk_out;
end architecture;
```

PR 5.14 Program 5.14

Example 5.9 Design a frequency divider that generates a clock frequency of 1 Hz from a clock frequency of 50 MHz.

Solution 5.9 We need a counter for the required clock frequency. Using the formula

$$\frac{f}{2K}$$

for the desired frequency, i.e.,

$$\frac{f}{2K} = 1$$

and substituting $f = 50,000,000$, we get $K = 25,000,000$. Thus, using a 1–24,999,999 counter, we can achieve the desired clock frequency as in PR 5.15.

```
library ieee;
use ieee.std_logic_1164.all;

entity clk_1Hz is
   port ( clk_50MHz: in std_logic;
          clk_1Hz: out std_logic);
end entity;

architecture logic_flow of clk_1Hz is
   signal count: natural range 1 to 25_000_000;
   signal temp_clk_out: std_logic;
begin
```

```
process(clk_50MHz)
   begin
      if (rising_edge(clk_50MHz)) then
         count <= count + 1;
         if(count= 25_000_000) then
            temp_clk_out<=not temp_clk_out;
            count<=1;
         end if;
      end if;
   end process;
   clk_1Hz <= temp_clk_out;
end architecture;
```

PR 5.15 Program 5.15

Example 5.10 Design a frequency divider that generates a clock frequency of 4 MHz from a clock frequency of 100 MHz.

Solution 5.10 Using the formula

$$\frac{f}{2K}$$

for the desired frequency, i.e.,

$$\frac{f}{2K} = 4,000,000$$

and substituting $f = 100,000,000$, we get $K = 12.5$. Thus, using a 0–11 counter, we can approximately achieve the desired clock frequency as in PR 5.16.

```
library ieee;
use ieee.std_logic_1164.all;

entity clk_4MHz is
   port ( clk_100MHz: in std_logic;
          clk_4MHz: out std_logic);
end entity;

architecture logic_flow of clk_4Hz is
   signal count: natural range 0 to 12;
   signal temp_clk_out: std_logic;
begin
```

```
process(clk_100MHz)
   begin
      if (rising_edge(clk_100MHz)) then
         count <= count + 1;
         if(count=12) then
            temp_clk_out<=not temp_clk_out;
            count<=0;
         end if;
      end if;
   end process;
   clk_4MHz <= temp_clk_out;
end architecture;
```

PR 5.16 Program 5.16

5.2.6 BCD to SS Converter with 1 s BCD Counter

Example 5.11 Design a digital circuit that displays the digits 0–9 every 1 s on seven segment display in a sequential manner for an FPGA device. Assume that FPGA has 100 MHz clock.

Solution 5.11 To display the digits 0–9 every 1 s on seven segment display in a sequential manner, we need a counter that increments its output every one second, and a seven segment display connected in series with counter. The black box representation of the circuit is depicted in Fig. 5.12.

Fig. 5.12 Digital circuit that displays the digits 0–9 every 1 s on seven segment display

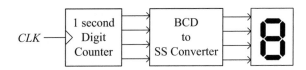

For the implementation of 1 s counter, we need a clock frequency of 1 Hz. The generation of 1 Hz frequency can be achieved using 1–49,999,999 counter as in PR 5.17.

PR 5.17 Program 5.17

```
library ieee;
use ieee.std_logic_1164.all;

entity clk_1Hz is
    port ( clk_100MHz: in std_logic );
end entity;

architecture logic_flow of clk_1Hz is
    signal count: positive range 1 to 50_000_000;
    signal clk_1Hz: std_logic;
begin

process(clk_100MHz)
    begin
    if (rising_edge(clk_100MHz)) then
        count <= count + 1;
        if(count=50_000_000) then
            clk_1Hz <=not clk_1Hz;
            count<=1;
        end if;
    end if;
end process;

end architecture;
```

Once we have the 1 Hz clock generator, we can add the seven segment display unit in a separate process as in PR 5.18.

PR 5.18 Program 5.18

```vhdl
library ieee;
use ieee.std_logic_1164.all;

entity clk_1Hz is
  port ( clk_100MHz: in std_logic;
         SSD: out std_logic_vector(6 downto 0));
end entity;

architecture logic_flow of clk_1Hz is
  signal count: natural range 1 to 50_000_000;
  signal digit: natural range 0 to 9;
  signal clk_1Hz: std_logic;
begin
process(clk_100MHz) % 1Hz  clock generation
 begin
  if (rising_edge(clk_100MHz)) then
    count <= count + 1;
    if(count=50_000_000) then
      clk_1Hz <=not clk_1Hz;
      count<=1;
    end if;
  end if;
 end process;
process(clk_1Hz) % SS_Display
 begin
    if(digit=9) then
       digit<=0;
    else
      digit<= digit+1;
    end if;
 end process;
   SSD<="0000001" when digit =0 else
        "1001111" when digit =1 else
        "0010010" when digit =2 else
        "0000110" when digit =3 else
        "1001100" when digit =4 else
        "0100100" when digit =5 else
        "0100000" when digit =6 else
        "0001111" when digit =7 else
        "0000000" when digit =8 else
        "0000100";
end architecture;
```

5.3 The Wait Statement

The **wait** statement can be used in some sequential program units. The use of the wait statement can be in one of the forms:

> **wait**;
> **wait until** conditional statement;
> **wait on** signal sensitivity list;
> **wait for** time duration.

We cannot use the different type of the **wait** statements:

> inside a process with a sensitivity list,
> inside a process which is called from a procedure having a sensitivity list,
> inside a function unit,
> inside a procedure which is called from a function.

The statements **wait for** and **wait** are used in simulation programs, i.e., in test-benches, and they are not synthesizable.

5.3.1 Wait Until

The statement **wait until** is mostly considered for the implementation of clocked sequential circuits.

Example 5.12 We can implement the D-type flip-flop using the **wait until** statement as in PR 5.19.

PR 5.19 Program 5.19

```
library ieee;
use ieee.std_logic_1164.all;

entity D_FlipFlop is
    port(d: in std_logic;
         clk: in std_logic;
         qa: out std_logic;
         qb: out std_logic);
end entity;
architecture logic_flow of D_FlipFlop is
begin
    process
    begin
        wait until (clk'event and clk='1');
        qa<=d;
        qb<=not d;
    end process;
end architecture;
```

We can use **wait until** statement more than once in a **process** unit. For instance, the process unit of the D-type flip-flop with enable input can be written as in PR 5.20.

PR 5.20 Program 5.20

```
process
    begin
        wait until (enable='1');
        wait until (clk'event and clk='1');
        qa<=d;
        qb<=not d;
end process;
```

5.3.2   Wait on

The **wait on** statement is used to check the changes in the signal parameter list, and if any change occurs in the parameter list, the process unit is executed. The processes P1 and P2 in PR 5.21 are equal to each other.

PR 5.21 Program 5.21

```
P1: process(clk, enable)
begin
  --vhdl statements
end process;

P2: process
begin
  -- vhdl  statements
  wait on clk, enable;
end process;
```

The position of the **wait on** statement inside the process unit is not an important issue. We can place it in any suitable place inside the **procedure** unit. However, we prefer to place it after the **begin**, or before the **end** statements.

Example 5.13 We can implement the D-type flip-flop using the **wait on** statement as in PR 5.22.

PR 5.22 Program 5.22

```
library ieee;
use ieee.std_logic_1164.all;

entity D_FlipFlop is
    port(d: in std_logic;
         clk: in std_logic;
         qa: out std_logic;
         qb: out std_logic);
end entity;
architecture logic_flow of D_FlipFlop is
begin
    process
    begin
        wait on clk;
        if(clk'event and clk='1') then
            qa<=d;
            qb<=not d;
        end if;
    end process;
end architecture;
```

5.4 Case Statement

Another statement used in sequential programming is the **case** statement. The syntax of the **case** statement is as follows:

```
case parameter is
when value1 => statement1;
when value2 => statement2;
         ⋮
end case;
```

Example 5.14 Implement the 4×1 multiplexer shown in Fig. 5.13 using the **case** sequential statement.

Fig. 5.13 4×1 multiplexer

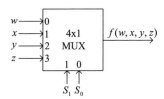

Solution 5.14 The implementation of 4×1 multiplexer using the case statement is given in PR 5.23.

PR 5.23 Program 5.23

```
library ieee;
use ieee.std_logic_1164.all;

entity multiplexer_4x1 is
  port( w, x, y, z: in std_logic;
        sel: in std_logic_vector(1 downto 0);
        f: out std_logic );
end entity;
architecture logic_flow of multiplexer_4x1 is
begin
  process(w, x, y, z, sel)
    begin
      case sel is
        when "00" => f<=w;
        when "01" => f<=x;
        when "10" => f<=y;
        when others => f<=z;
      end case;
    end process;
end architecture;
```

We can consider more than a single value at the same line of the **case** statement. In this case the syntax of the **case** statement becomes as:

```
case parameter is
    when value11 | value12 | value13 => statement1;
    when value21 | value22 | value23 => statement2;
                        ⋮
end case;
```

Example 5.15 In PR 5.24, the use of **case** statement for multiple parameter values is illustrated.

PR 5.24 Program 5.24

```
case num is
    when 21 | 34 => f<=w;
    when 45 | 57 => f<=x;
    when others  => f<=z;
end case;
```

5.5 Loop Statements

Loop structures are vital part of the sequential programming. There are three kinds of loop structures available in VHDL programming. These structures are:

[label:] **loop**
          program statements
    end loop [label] ;

[label:] **for** parameter **in** range **loop**
          program statements
    end loop [label] ;

[label:] **while** conditional statement **loop**
          program statements
    end loop [label] ;

where labels are optional and they can be omitted.
Sequential loops can also be utilized with **next** and **exit** statements.

5.5.1 Next and Exit Statements

Next statement is used if we want to continue with the next iteration without finalizing the current iteration. **Exit** statement is used to exit the current loop. The syntax of the **next** and **exit** statements are as follows:

[label :] **next** [loop_label] [**when** condition];
[label :] **exit** [loop_label] [**when** condition];

Example 5.16 Possible uses of **next** and **exit** statements are given in PR 5.25.

PR 5.25 Program 5.25

```
next;
next outer_loop;
next when x<=y;
next inner_loop when x and y='1'

exit;
exit outer_loop;
exit when inp xor pass="0000";
exit inner_loop when inp xor pass="0000";
```

Example 5.17 In PR 5.26, we provide small examples for the three kind of loops used in sequential programming.

```
my_loop: loop
            if (my_array(index)=1) then
                one_counter<=one_counter+1;
            end if;
                index:=index+1;
            exit my_loop when one_counter >= 16;
            end loop;
```

```
initialize_my_array: for index in 0 to 15 loop
                my_array(index) := 0;
            end loop;
```

```
while_loop: while index <= 8 loop
                z(index) <= x(index) xor y(index);
                index:= index + 1;
            end loop while_loop;
```

PR 5.26 Program 5.26

5.6 Example Sequential Logic Circuit Implementation in VHDL

In this section, we will provide examples for the implementation of sequential logic circuits that can involve combinational logic circuits.

Example 5.18 Implement the digital circuit of Fig. 5.14 in VHDL language.

Fig. 5.14 Digital circuit for Example 5.18

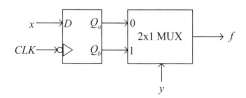

Solution 5.18 Let's label the internal signals by q_a and q_b as in Fig. 5.15.

Fig. 5.15 Digital circuit with internal signals labeled

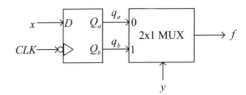

Considering Fig. 5.15, we can write the title part and declare the internal signals in the declarative part of the architecture as in PR 5.27.

```
library ieee;
use ieee.std_logic_1164.all;

entity logic_circuit is
    port(x, y, clk: in std_logic;
             f: out std_logic );
end entity;
```

```
architecture logic_flow of logic_circuit is
    signal qa, qb: std_logic;
begin
end architecture;
```

PR 5.27   Program 5.27

In the next step, we implement the operation of the circuit in the body part of the architecture as in PR 5.28.

PR 5.28 Program 5.28

```
library ieee;
use ieee.std_logic_1164.all;

entity logic_circuit is
    port(x, y, clk: in std_logic;
            f: out std_logic );
end entity;

architecture logic_flow of logic_circuit is
    signal qa, qb: std_logic;
begin
    process(clk)
    begin
       if (falling_edge(clk)) then
                qa<=x;   qb<=not x;
          end if;
       end process;

       f<=qa when y='0' else
          qb;
end architecture;
```

Example 5.19 Implement the digital circuit in Fig. 5.16 using VHDL.

Fig. 5.16 Digital circuit for
Example 5.19

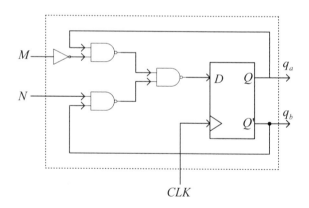

CLK

Solution 5.19 We first label the internal signals of the circuit as shown in Fig. 5.17 and implement the I/O ports in entity unit, and declare the internal signals in the declarative part of the architecture unit as in PR 5.29.

Fig. 5.17 Digital circuit with internal signals labeled

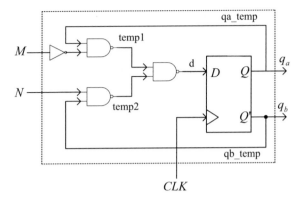

PR 5.29 Program 5.29

```
library ieee;
use ieee.std_logic_1164.all;

entity logic_circuit is
    port(m, n, clk: in std_logic;
             qa, qb: out std_logic );
end entity;

architecture logic_flow of logic_circuit is
    signal d, qa_temp, qb_temp: std_logic;
    signal temp1, temp2: std_logic;
begin

end architecture;
```

From Fig. 5.17, we can write the mathematical expressions for the internal signals as

$$temp1 = (qa\_temp \cdot M')' \quad temp2 = (qb\_temp \cdot N)' \quad d = temp1 \cdot temp2$$

which are implemented together with the operation of D-type flip-flop in the body part of the architecture unit as in PR 5.30.

PR 5.30 Program 5.30

```
library ieee;
use ieee.std_logic_1164.all;

entity logic_circuit is
    port(m, n, clk: in std_logic;
         qa, qb: out std_logic );
end entity;

architecture logic_flow of logic_circuit is
    signal d, qa_temp, qb_temp: std_logic;
    signal temp1, temp2: std_logic;
begin
    temp1<= (not m) nand qa_temp;
    temp2<= n nand qb_temp;
    d<=temp1 nand temp2;
    process(clk)
    begin
      if (rising_edge(clk)) then
        qa_temp<=d;
        qb_temp <=not d;
      end if;
    end process;
  qa<= qa_temp;
  qb<= qb_temp;
end architecture;
```

Exercise: Implement the digital circuit in Fig. 5.18 using VHDL.

Fig. 5.18 Digital circuit for exercise

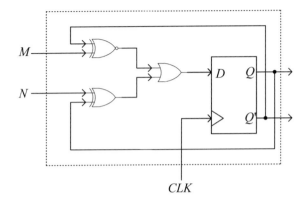

Example 5.20 Implement the digital circuit in Fig. 5.19 using VHDL.

Fig. 5.19 Digital circuit for Example 5.20

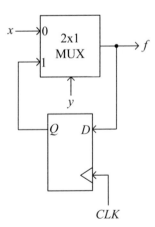

Solution 5.20 We first label the internal signals as in Fig. 5.20.

Fig. 5.20 Digital circuit with
internal signals labeled

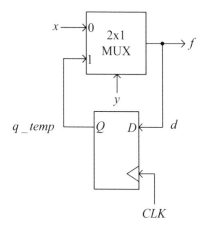

Considering Fig. 5.20, we define the port parameters in the entity unit and
internal signals in the body part of the architecture unit as in PR 5.31.

PR 5.31 Program 5.31

```
library ieee;
use ieee.std_logic_1164.all;

entity logic_circuit is
     port(x, y, clk: in std_logic;
          f: buffer std_logic );
end entity;

architecture logic_flow of logic_circuit is
     signal d, q_temp: std_logic;
begin

end architecture;
```

The operation of the combinational circuit, i.e., multiplexer unit, can be
implemented in the declarative part of the architecture unit as in PR 5.32.

PR 5.32 Program 5.32

```
library ieee;
use ieee.std_logic_1164.all;

entity logic_circuit is
    port(x, y, clk: in std_logic;
            f: buffer std_logic );
end entity;
```

```
architecture logic_flow of logic_circuit is
    signal d, q_temp: std_logic;
begin
    f<=x when y='0' else
        q_temp;
    d<=f;
end architecture;
```

Adding the implementation of the D flip flop, we get the full VHDL program as in PR 5.33.

PR 5.33 Program 5.33

```
library ieee;
use ieee.std_logic_1164.all;

entity logic_circuit is
    port(x, y, clk: in std_logic;
            f: buffer std_logic );
end entity;

architecture logic_flow of logic_circuit is
    signal d, q_temp: std_logic;
begin
    f<=x when y='0' else
        q_temp;
    d<=f;
    process(clk)
    begin
        if (rising_edge(clk)) then
            q_temp<=d;
        end if;
    end process;
end architecture;
```

Example 5.21 Implement the digital circuit in Fig. 5.21 using VHDL.

Solution 5.21 The digital circuit shown in Fig. 5.21 can be implemented in a number of ways. One implementation of the circuit shown in Fig. 5.21 is given in PR 5.34.

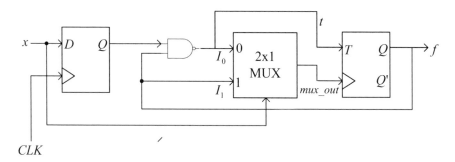

Fig. 5.21 Digital circuit for Example 5.21

```
library ieee;
use ieee.std_logic_1164.all;

entity logic_circuit is
    port(x, clk: in std_logic;
         f: buffer std_logic );
end entity;

architecture logic_flow of logic_circuit is
    signal d, qd_temp, t: std_logic:='0';
    signal i0, i1, mux_out: std_logic;
    signal qt_temp: std_logic;
begin
    mux_out <=i0 when x='0' else
              i1;
    i0<= qd_temp nand f;
    i1<= f;
    t<=i0;
    process(mux_out)
    begin
      if (rising_edge(mux_out)) then
        if(t='1') then
          qt_temp<=not qt_temp;
        else
          qt_temp<=qt_temp;
        end if;
      end if;
    end process;
    f<=qt_temp;
```

```
process(clk)
begin
  if (rising_edge(clk)) then
    qd_temp<=x;
  else
    qt_temp<=qt_temp;
  end if;
end process;
end architecture;
```

PR 5.34   Program 5.34

Exercise: Implement the digital circuit in Fig. 5.22 using VHDL.

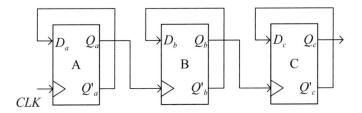

Fig. 5.22 Digital circuit for VHDL implementation

Exercise: Implement the digital circuit in Fig. 5.23 using VHDL.

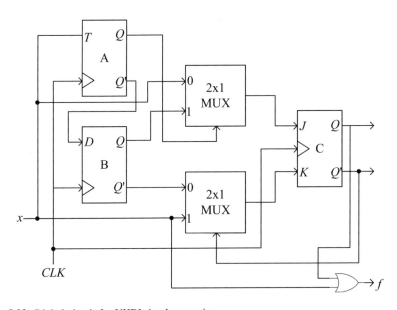

Fig. 5.23 Digital circuit for VHDL implementation

Problems

(1) Using 100 MHz FPGA clock frequency design a 1 kHz clock generator.
(2) Assume that FPGA device has 100 MHz clock generator. Design two clock sources with 1 kHz clock frequency such that if the first clock waveform is denoted by $clk(t)$, then the second waveform is denoted by $clk(t + T/2)$ where T is the period of $clk(t)$, i.e., the second clock waveform has some phase shift w.r.t. first clock waveform.

(3) Assume that you have two clock waveforms with the same frequency F_c, and
     if the first clock waveform is denoted by $clk(t)$, then the second waveform is
     denoted by $clk(t+T/2)$ where T is the period of $clk(t)$. Using these two
     waveforms design a clock generator with frequency $2F_c$.

(4) Write a VHDL program such that the digits 0–9 are displayed on seven
     segment display with 1 s time durations.

(5) The characteristic table of the *MN* flip flop is displayed on the left of Fig. 5.P1.
     Implement the logic circuit shown on the right of Fig. 5.P1 in VHDL. Do not
     use components in your implementation.

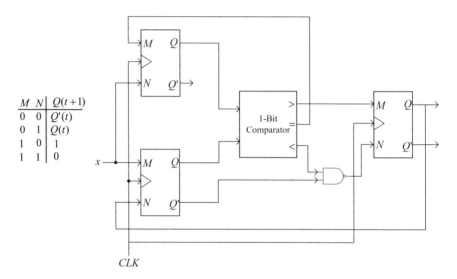

P5.1 Digital circuit for VHDL implementation

(6) Using a variable object in your frequency divider, design 1 kHz clock gen-
     erator using 100 MHz FPGA clock generator.

(7) Implement the circuit shown in Fig. 5.P2 in VHDL. Use components in your
     implementation.

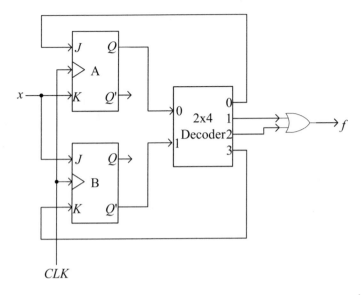

P5.2 Digital circuit for VHDL implementation

(8) Implement the circuit shown in Fig. 5.P3 in VHDL.

P5.3 Digital circuit for
VHDL implementation

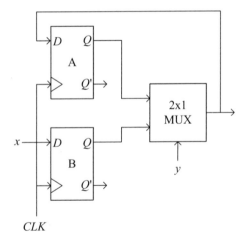

(9) Implement the circuit shown in Fig. 5.P4 in VHDL. In your implementation
do not use any concurrent coding for logic units, i.e., use only **process** units
for the implementation of logic blocks.

P5.4 Digital circuit for
VHDL implementation

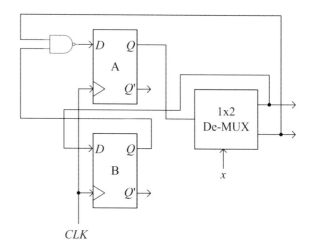

(10) Implement the circuit shown in Fig. 5.P5 in VHDL.

5.P5 Digital circuit for
VHDL implementation

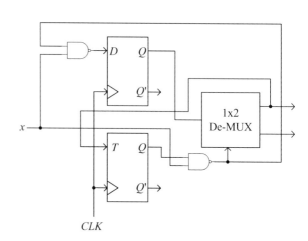

(11) Write a VHDL program that displays the numbers $00, 01, 02, \ldots 59, 60$ in a
       sequential manner with 1 s durations each on two seven-segment displays.
(12) Implement a digital clock in VHDL.
(13) Implement the circuit shown in Fig. 5.P6 in VHDL.
(14) Write a VHDL program that displays the sequence 7, 21, 45, 32, 78, 12, 98,
       48, 87, 2 repeatedly on two seven-segment displays, and elements of the
       sequence are displayed in sequential manner with 1 s durations for each
       number on two seven-segment displays.

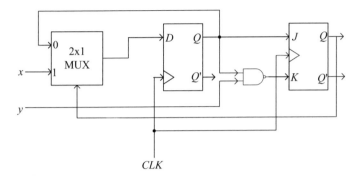

P5.6 Digital circuit for VHDL implementation

(15) Assume that FPGA device has 50 MHz clock generator. Design four clock sources with 1 kHz clock frequency such that if the first clock waveform is denoted by $clk(t)$, then the second, third and fourth waveforms are denoted by $clk(t+T/4)$, $clk(t+2T/4)$, $clk(t+3T/4)$ where T is the period of $clk(t)$.

(16) If the clock frequency of the FPGA is 100 MHz, can you design a clock generator with frequency 200 MHz? If yes, implement your design in VHDL.

Chapter 6
VHDL Implementation of Logic Circuits Involving Registers and Counters

In this chapter we will explain the implementation of logic circuits involving registers and counters. In the previous chapter we focused on the implementation of simpler units, such as flip-flops, however, in this chapter, we will solve exercises for the implementation of more complex logic circuits.

6.1 Shift Registers

Shift registers are constructed from flip-flops. Such type of registers is usually employed in serial communication operations. They can also be used in arithmetic units employing serial addition, subtraction, multiplication, etc. In this section we will consider logic circuits employing shift registers.

Example 6.1 Four-bit shift register is shown in Fig. 6.1. Implement the 4-bit shift register shown in Fig. 6.1 in VHDL.

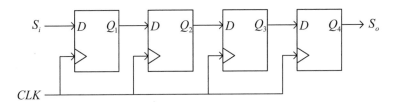

Fig. 6.1 Shift register with D-type flip-flops

© Springer Nature Singapore Pte Ltd. 2019
O. Gazi, *A Tutorial Introduction to VHDL Programming*,
https://doi.org/10.1007/978-981-13-2309-6_6

Solution 6.1 The internal signals of the shift register in Fig. 6.1 are explicitly depicted in Fig. 6.2.

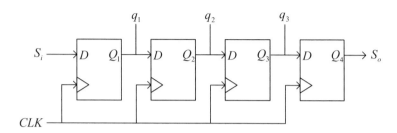

Fig. 6.2 Shift register with internal signals labeled

Considering Fig. 6.2, we can define the internal signals in the declarative part of the architecture unit and input-output ports in the entity part as in PR 6.1.

PR 6.1 Program 6.1

```
library ieee;
use ieee.std_logic_1164.all;

entity shift_register is
    port( si, clk: in std_logic;
            so: out std_logic );
end entity;

architecture logic_flow of shift_register is
    signal q1, q2, q3: std_logic;
begin

end architecture;
```

Considering the operations of the D flip-flops, we can write a procedure for the operation of the shift register in the body of the architecture unit as in PR 6.2.

PR 6.2 Program 6.2

```
library ieee;
use ieee.std_logic_1164.all;

entity shift_register is
    port( si, clk: in std_logic;
            so: out std_logic );
end entity;

architecture logic_flow of shift_register is
    signal q1, q2, q3: std_logic;
begin
  process(clk)
    begin
        if (clk'event and clk='1') then
          so<=q3;
          q3<=q2;
          q2<=q1;
          q1<=si;
        end if;
      end process;
end architecture;
```

We can implement the shift register in Fig. 6.1 in an alternative way. In this implementation, we declare the internal signals using **std_logic_vector** as in PR 6.3.

PR 6.3 Program 6.3

```
library ieee;
use ieee.std_logic_1164.all;

entity shift_register is
    port( si, clk: in std_logic;
            so: out std_logic );
end entity;

architecture logic_flow of shift_register is
    signal dq_array: std_logic_vector(3 downto 0);
begin

end architecture;
```

The operation of the shift register can be written using the internal signal logic vector as in PR 6.4.

PR 6.4 Program 6.4

```
library ieee;
use ieee.std_logic_1164.all;

entity shift_register is
    port( si, clk: in std_logic;
            so: out std_logic );
end entity;

architecture logic_flow of shift_register is
  signal dq_array: std_logic_vector(3 downto 0);
begin

  process(clk)
    begin
      if (clk'event and clk='1') then
        dq_array<=si & dq_array(3 downto 1);
      end if;
    end process;
  so<=dq_array(0);
end architecture;
```

6.1.1 Shift Register with Parallel Load

Shift registers with parallel load capability have the ability of loading external data to the binary cells in a parallel manner.

Example 6.2 Implement the 4-bit shift register with parallel load property shown in Fig. 6.3 using VHDL.

Fig. 6.3 Shift register with parallel load property

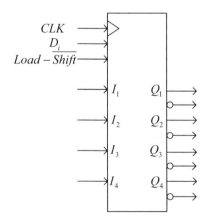

Solution 6.2 The circuit shown in Fig. 6.3 has both serial shifting and parallel load capability. The outputs are directly taken from the internal signals of the shift registers. Considering this issue, we can define the port and internal signals as in PR 6.5.

PR 6.5 Program 6.5

```
library ieee;
use ieee.std_logic_1164.all;

entity shift_register is
    port( di, clk, load_shift: in std_logic;
          inp: in std_logic_vector(4 downto 1);
          q: out std_logic_vector(4 downto 1);
          qc: out std_logic_vector(4 downto 1) );
end entity;

architecture logic_flow of shift_register is

  signal dq_array: std_logic_vector(4 downto 1);

begin
```

The parallel load and shifting operations of the circuit can be achieved using the process in PR 6.6.

PR 6.6 Program 6.6

```
architecture logic_flow of shift_register is
  signal dq_array: std_logic_vector(4 downto 1);
begin
  process(clk)
    begin
      if(load_shift='1') then
        dq_array <= inp;
      elsif (clk'event and clk='1') then
        dq_array<=di & dq_array(4 downto 2);
      end if;
    end process;
end architecture;
```

Finally, equating the outputs to the internal signals after process unit in the body part of the architecture, we get the complete program as in PR 6.7.

PR 6.7 Program 6.7

```
library ieee;
use ieee.std_logic_1164.all;

entity shift_register is
    port( di, clk, load_shift: in std_logic;
            inp: in std_logic_vector(4 downto 1);
            q: out std_logic_vector(4 downto 1);
            qc: out std_logic_vector(4 downto 1) );
end entity;

architecture logic_flow of shift_register is
  signal dq_array: std_logic_vector(4 downto 1);
begin
  process(clk)
      begin
          if(load_shift='1') then
            dq_array <= inp;
          elsif (clk'event and clk='1') then
            dq_array<=di & dq_array(4 downto 2);
          end if;
        end process;
        q<= dq_array;
        qc<=not dq_array;
end architecture;
```

6.1.2   Logic Circuits Involving Shift Registers and Counter

In this subsection we will consider the implementation of logic circuits involving clocked and combinational logic units.

Example 6.3 Implement the digital circuit of Fig. 6.4 in VHDL. The circuit contains two 4-bit shift register, one multiplexer, and one 1-bit counter.

Fig. 6.4 Digital circuit for example 6.3

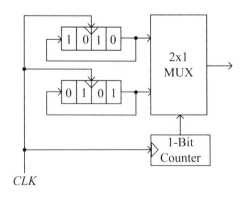

Solution 6.3 The circuit in Fig. 6.4 has one input and one output. The input is taken from clock source and the output is take from the multiplexer output. The contents of shift registers and output of the 1-bit counter can be considered as internal signals. Then, we can define the port signals and internal signals as in PR 6.8.

```
library ieee;
use ieee.std_logic_1164.all;

entity logic_circuit is
      port(clk, initialise: in std_logic;
            q: out std_logic );
end entity;

architecture logic_flow of logic_circuit is
   signal shift_reg1: std_logic_vector(3 downto 0):="1010";
   signal shift_reg2: std_logic_vector(3 downto 0):="0101";
   signal counter: natural range 0 to 1:=0;
begin
end architecture:
```

PR 6.8 Program 6.8

Although there is no initialization control signal in the circuit, it is logical to define one as an input port as in PR 6.8. Since, the signal initializations in the declarative part of the architecture does not work in hardware implementation. They are valid only for software simulation. For this reason, we need to define a control signal for signal initialization as in PR 6.8. The operation of the counter and initialization process can be achieved using the **process** unit in PR 6.9.

```
process(clk, initialise)
begin
   if (clk'event and clk='1') then
      if(initialise='1') then
         shift_reg1<="1010";
         shift_reg2<="0101";
         counter<=0;
      end if;
      if(counter=1) then
         counter<=0;
      else
         counter<=counter+1;
      end if;
   end if;
end process;
```

PR 6.9 Program 6.9

The two shift registers can be implemented using two separate process units as in PR 6.10.

```
process(clk)
begin
   if (clk'event and clk='1') then
       shift_reg1<=shift_reg1(0) & shift_reg1(3 downto 1);
   end if;
end process;

process(clk)
begin
  if (clk'event and clk='1') then
       shift_reg2<=shift_reg2(0) & shift_reg2(3 downto 1);
  end if;
end process;
```

PR 6.10 Program 6.10

Finally, port output values can be obtained from the shift registers in the body part of the architecture after shift register implementations as in PR 6.11.

```
process(clk)
begin
   if (clk'event and clk='1') then
       shift_reg1<=shift_reg1(0) & shift_reg1(3 downto 1);
   end if;
end process;
```

```
process(clk)
begin
  if (clk'event and clk='1') then
       shift_reg2<=shift_reg2(0) & shift_reg2(3 downto 1);
  end if;
end process;

q<=shift_reg1(0) when counter=0 else
    shift_reg2(0);
end architecture;
```

PR 6.11 Program 6.11

Unifying all the program segments we wrote up to know, we get the full program as in PR 6.12.

```
library ieee;
use ieee.std_logic_1164.all;

entity logic_circuit is
    port(clk, initialise: in std_logic;
         q: out std_logic );
end entity;

architecture logic_flow of logic_circuit is
  signal shift_reg1: std_logic_vector(3 downto 0):="1111";
  signal shift_reg2: std_logic_vector(3 downto 0):="0101";
  signal counter: natural range 0 to 1:=0;
begin
process(clk, initialise)
begin
   if (clk'event and clk='1') then
      if(initialise='1') then
         shift_reg1<="1010";
         shift_reg2<="0101";
         counter<=0;
      end if;
      if(counter=1) then
         counter<=0;
       else
          counter<=counter+1;
       end if;
    end if;
end process;
process(clk)
begin
   if (clk'event and clk='1') then
       shift_reg1<=shift_reg1(0) & shift_reg1(3 downto 1);
    end if;
end process;

process(clk)
begin
  if (clk'event and clk='1') then
       shift_reg2<=shift_reg2(0) & shift_reg2(3 downto 1);
   end if;
end process;
q<=shift_reg1(0) when counter=0 else
    shift_reg2(0);
end architecture;
```

PR 6.12   Program 6.12

Exercise Implement the circuit of Fig. 6.5, where there are three shift registers and one 2×1 multiplexer, in VHDL.

Fig. 6.5 Digital circuit for exercise

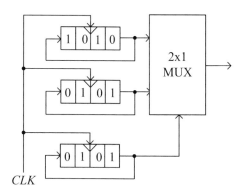

6.1.3   Serial Transfer Unit

Serial transfer units are used to transfer the contents of one shift register to a second shift register. The logic diagram of the serial transfer unit is depicted in Fig. 6.6.

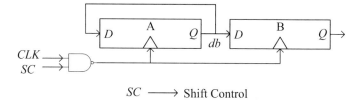

Fig. 6.6 Serial transfer unit

Example 6.4 Implement the serial transfer unit shown in Fig. 6.6 in VHDL. Assume that shift registers contain 8 bits.

Solution 6.4 The serial transfer unit has two input signals and one output signal. The contents of the shift registers and output of the first shift register can be considered as internal signals. Considering this information, we can implement the ports and internal signals as in PR 6.13.

```
library ieee;
use ieee.std_logic_1164.all;

entity logic_circuit is
    port(clk, sc: in std_logic;
            qb: out std_logic );
end entity;

architecture logic_flow of logic_circuit is
  signal shift_reg_A: std_logic_vector(7 downto 0);
  signal shift_reg_B: std_logic_vector(7 downto 0);
  signal db: std_logic;
begin

end architecture;
```

PR 6.13 Program 6.13

The two shift registers of Fig. 6.6 can be implemented using two **process** units as in PR 6.14.

```
process(sc,clk)
   begin
   if(sc='1') then
      if (clk'event and clk='0') then
         shift_reg_A<=shift_reg_A(0) & shift_reg_A(7 downto 1);
      end if;
   end if;
end process;

process(sc,clk)
   begin
   if(sc='1') then
      if (clk'event and clk='0') then
         shift_reg_B<=db & shift_reg_B(7 downto 1);
      end if;
   end if;
end process;
```

PR 6.14 Program 6.14

Finally, outputs of the first and second shift registers can be implemented in the declarative part of the architecture unit using the codes in PR 6.15

PR 6.15 Program 6.15

```
db<=shift_reg_A(0);
qb<=shift_reg_B(0);
```

Overall, our complete program becomes as in PR 6.16.

```
library ieee;
use ieee.std_logic_1164.all;

entity logic_circuit is
    port(clk, sc: in std_logic;
            qb: out std_logic );
end entity;

architecture logic_flow of logic_circuit is
  signal shift_reg_A: std_logic_vector(7 downto 0);
  signal shift_reg_B: std_logic_vector(7 downto 0);
  signal db: std_logic;
begin

    db<=shift_reg_A(0);
    qb<=shift_reg_B(0);

  process(sc,clk)
    begin
     if(sc='1') then
        if (clk'event and clk='0') then
          shift_reg_A<=shift_reg_A(0) & shift_reg_A(7 downto 1);
        end if;
     end if;
    end process;

  process(sc,clk)
    begin
     if(sc='1') then
        if (clk'event and clk='0') then
          shift_reg_B<=db & shift_reg_B(7 downto 1);
        end if;
     end if;
    end process;

end architecture;
```

PR 6.16 Program 6.16

6.1.4   Serial Adder Unit

The serial adder is used to sum the contents of two shift registers, and it is depicted in Fig. 6.7.

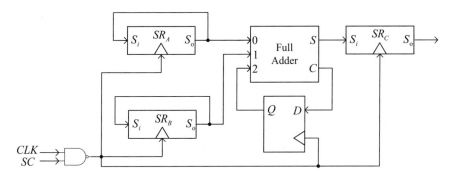

Fig. 6.7   Serial adder unit

Example 6.5 Implement the serial adder shown in Fig. 6.7 in VHDL.

Solution 6.5 First, let's label the internal signals as in Fig. 6.8.

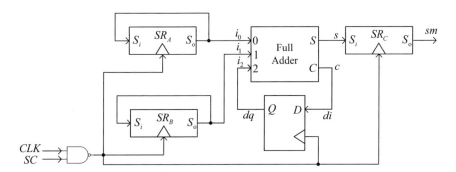

Fig. 6.8   Serial adder unit with internal signals labeled

Considering Fig. 6.8, we can declare the input output ports and internal signals in the entity and declarative part of the architecture unit as in PR 6.17.

```
library ieee;
use ieee.std_logic_1164.all;

entity logic_circuit is
     port(clk, sc: in std_logic;
           sm: out std_logic );
end entity;

architecture logic_flow of logic_circuit is

  signal shift_reg_A: std_logic_vector(7 downto 0);
  signal shift_reg_B: std_logic_vector(7 downto 0);
  signal shift_reg_C: std_logic_vector(7 downto 0);
  signal dq, di: std_logic;
  signal i0, i1, i2: std_logic; -- Inputs of full adder
  signal s, c: std_logic; -- Outputs of full adder

begin

end architecture;
```

PR 6.17 Program 6.17

In the next step, we connect the internal signals to each other in the body of the architecture unit as in PR 6.18.

PR 6.18 Program 6.18

```
begin

     i0<=shift_reg_A(0);
     i1<=shift_reg_B(0);
     i2<= dq;

     s<= (i0 xor i1) xor i2;
     c<=(i0 and i1) or (i0 and i2) or (i1 and i2) ;

     sm<= shift_reg_C(0);

end architecture;
```

Lastly, we implement the operations of clocked sequential circuits of Fig. 6.7 using processes as in PR 6.19.

```
begin
    i0<=shift_reg_A(0);
    i1<=shift_reg_B(0);
    i2<= dq;
    s<= (i0 xor i1) xor i2;
    c<=(i0 and i1) or (i0 and i2) or (i1 and i2) ;
    sm<= shift_reg_C(0);

SHRA: process(sc,clk)
        begin
        if(sc='1') then
          if (clk'event and clk='1') then
          shift_reg_A<=shift_reg_A(0) & shift_reg_A(7 downto 1);
          end if;
          end if;
        end process;

SHRB: process(sc,clk)
        begin
        if(sc='1') then
          if (clk'event and clk='1') then
            shift_reg_B<=shift_reg_B(0) & shift_reg_B(7 downto 1);
          end if;
          end if;
        end process;

SHRC: process(sc,clk)
        begin
        if(sc='1') then
          if (clk'event and clk='1') then
            shift_reg_C<=s & shift_reg_C(7 downto 1);
          end if;
          end if;
        end process;

DHFF: process(sc,clk)
        begin
        if(sc='1') then
          if (clk'event and clk='1') then
            dq<=c;
          end if;
          end if;
        end process;
end architecture;
```

PR 6.19 Program 6.19

Problems

(1) The circuit in Fig. 6.P1 performs the summation of three binary numbers. Implement the circuit in VHDL.

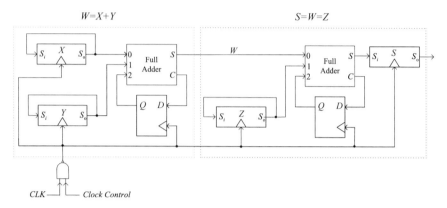

Fig. 6.P1 Digital circuit for the summation of three binary numbers

(2) The serial subtractor circuit is depicted in Fig. 6.P2. Implement the serial subtractor in VHDL.

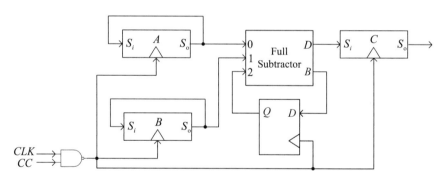

Fig. 6.P2 Serial subtractor

(3) The serial adder and subtractor circuit is shown in Fig. 6.P3. Implement this circuit in VHDL.

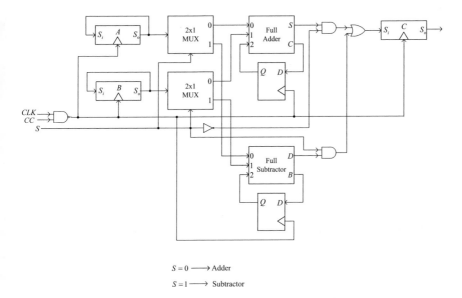

$S = 0 \longrightarrow$ Adder

$S = 1 \longrightarrow$ Subtractor

Fig. 6.P3 Serial adder and subtractor circuit

(4) Serial equality checker is shown in Fig. 6.P4. Implement this circuit in VHDL.

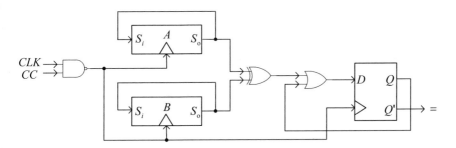

Fig. 6.P4 Serial equality checker

(5) Serial magnitude comparator circuit is shown in Fig. 6.P5. Implement this circuit in VHDL.

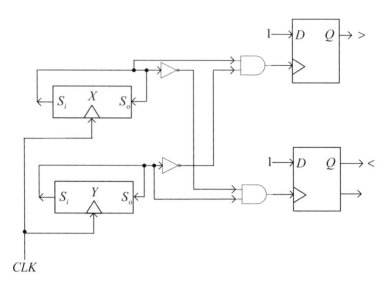

Fig. 6.P5 Serial magnitude comparator

(6) Implement the circuit in Fig. 6.P6 in VHDL.

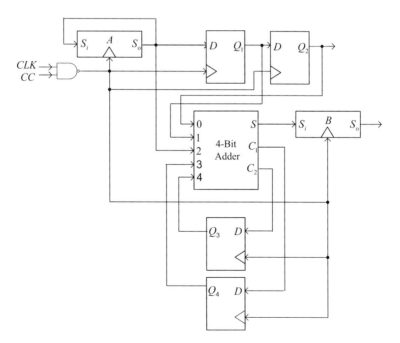

Fig. 6.P6 Digital circuit for VHDL implementation

(7) Implement the asynchronous up counter shown in Fig. 6.P7 in VHDL.

Fig. 6.P7 4-bit
asynchronous up counter

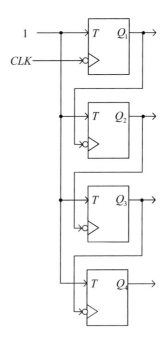

(8) Implement the asynchronous down counter shown in Fig. 6.P8 in VHDL.

Fig. 6.P8 4-bit
asynchronous down counter

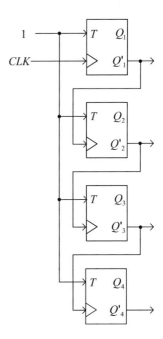

(9) Implement the synchronous BCD counter shown in Fig. 6.P9 in VHDL.

Fig. 6.P9 Synchronous BCD counter with D flip-flops

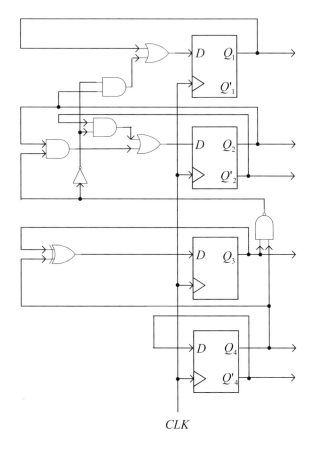

(10)   Implement the asynchronous BCD counter shown in Fig. 6.P10 in VHDL.

Fig. 6.P10 Asynchronous
BCD counter with D flip-flops

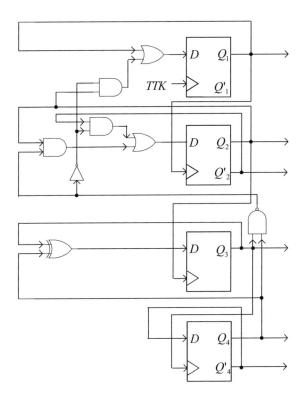

(11) Implement the asynchronous BCD counter shown in Fig. 6.P11 in VHDL.

Fig. 6.P11 Asynchronous
4-bit BCD counter with
JK flip-flops

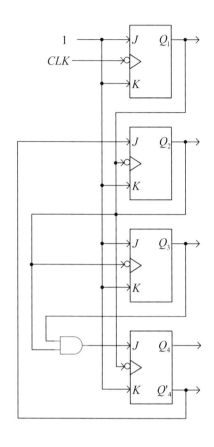

(12)   Implement the asynchronous Johnson counter shown in Fig. 6.P12 in VHDL.

Fig. 6.P12 Asynchronous
4-bit Johnson counter with
D flip-flops

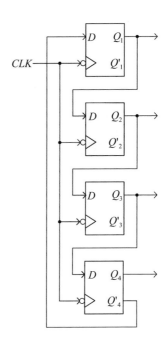

(13)   Implement the modulus counter shown in Fig. 6.P13 in VHDL.

Fig. 6.P13 Modulus counter

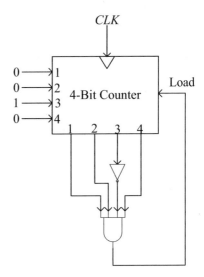

(14)   Implement the modulus counter shown in Fig. 6.P14 in VHDL.

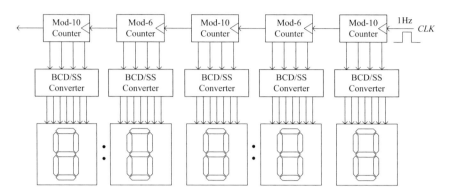

Fig. 6.P14 Digital circuit for VHDL implementation

Chapter 7
Packages, Components, Functions and Procedures

In this chapter packages, components, functions and procedures will be explained in details. The repeatedly used parts of program units, such as functions, procedures, components, and data declarations, such as constants and user defined data type declarations are placed into the package unit. In this way, the program units placed inside the package becomes portable, and less space is occupied in the main program window. Packages can be written as a separate program unit or they can be placed before the main program.

7.1 Packages

Packages consist of two sections, one is the declaration part of the package, the other is the body of the package. The declaration part usually contains the data declarations, and function or procedure headers, i.e., declarations. The body part of the package usually contains the implementation part of the functions or procedures. The syntax of the package is shown in PR 7.1. To use the package in our program, we need to include the package file at the header of the main program.

PR 7.1 Program 7.1

```
package package_name is
   declarative part
end package_name;

package body package_name is
   body part
end package  name;
```

© Springer Nature Singapore Pte Ltd. 2019   189
O. Gazi, *A Tutorial Introduction to VHDL Programming*,
https://doi.org/10.1007/978-981-13-2309-6_7

This is achieved using the statement

use work.package_name.**all**

Example 7.1 Package declaration can be made before the main program, i.e., main program and package declaration exists in the same file. In PR 7.2, package declaration is made before the inclusion of **ieee** library and package.

```
package my_package is
    constant  data_length: positive range 1 to 100:=32;
end my_package;

package body my_package is
    -- No statement yet
end my_package;

-- Package declarations must be made before the inclusion of other library packages

library ieee;
use ieee.std_logic_1164.all;

use work.my_package.all; -- Necessary to use the package

entity logic_circuit  is
    port( d: in std_logic_vector(data_length-1 downto 0);
          q: out std_logic_vector(data_length-1 downto 0) );
end entity;

architecture logic_flow of logic_circuit is
begin
    q<=d;
end architecture;
```

PR 7.2 Program 7.2

Usually package declarations should be made before the inclusion of any other library or packages. However, this rule may not be valid for some of the newly designed VHDL development platforms of some vendors.

Example 7.2 Package declaration and main program can exist in different files. The previous example can be implemented in two separate files as in PR 7.3 and PR 7.4.

my_package.vhd

```
package my_package is
    constant data_length: positive range 1 to 100:=32;
end my_package;

package body my_package is
    -- No statement yet
end my_package;
```

PR 7.3 Program 7.3

logic_circuit .vhd

```
library ieee;
use ieee.std_logic_1164.all;
use work.my_package.all; -- Necessary to use the package

entity logic_circuit is
    port( d: in std_logic_vector(data_length-1 downto 0);
          q: out std_logic_vector(data_length-1 downto 0) );
end entity;

architecture logic_flow of logic_circuit is
begin
    q<=d;
end architecture;
```

PR 7.4 Program 7.4

7.2   Components

If some of the logic units, such as, multiplexers, adders, flip flops etc., are repeatedly used in logic circuits, we can write a library for these units and use them whenever they are needed without over and over again re-implementing these units. This is achieved via the use of **component** utility.

If we had already implemented some of the logic units, and want to use these units in some other VHDL programs, we can declare these units as components. Components declarations are either done in package unit, or in the declarative part of the architecture. Once components are declared, we can instantiate them in any number in our main program.

Component declaration is the same as **entity** declaration except for the use of **component** keyword instead of **entity** keyword. The declaration and instantiation syntax of the components is as in PR 7.5.

PR 7.5 Program 7.5

> **component** component_name **is**
>   **port**(port declarations);
> **end component**;
>
> compLabel: component_name **port** **map**(port list);

Let's illustrate the subject with an example.

Example 7.3 4-bit shift register is shown in Fig. 7.1. Implement the 4-bit shift register shown in Fig. 7.1 using components.

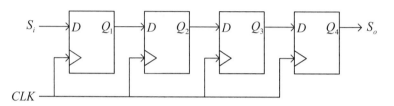

Fig. 7.1 Shift register for VHDL implementation

Solution 7.3 If Fig. 7.1 is inspected, we see that D flip-flop is repeatedly used. First, we implement the D flip-lop as in PR 7.6.

PR 7.6 Program 7.6

d ff.vhd

```
library ieee;
use ieee.std_logic_1164.all;

entity dff is
    port( d, clk: in std_logic;
          q: out std_logic );
end entity;

architecture logic_flow of dff is
begin
  process(clk)
    begin
      if (clk'event and clk='1') then
          q<=d;
      end if;
    end process;
end architecture;
```

In the next step, we have two ways of implementing the circuit in Fig. 7.1.

First Method: In the first method, create a package unit and declare the component in the package unit, and include the packet in main program where components are instantiated. The package unit where component declaration is made is depicted in PR 7.7.

PR 7.7 Program 7.7

my_comp_package.vhd

```
package my_comp_package is
component dff is
    port( d, clk: in std_logic;
            q: out std_logic );
    end component;
end my_comp_package;
```

Once, we declare the components, we can use them in any number in our main program. For this purpose, we include the packet in the header of our program as in PR 7.8 and declare the internal signals in the declarative part of the architecture unit.

PR 7.8 Program 7.8

```
library ieee;
use ieee.std_logic_1164.all;
use work. my_comp_package.all;

entity shift_register is
    port( si, clk: in std_logic;
            so: out std_logic );
end entity;

architecture logic_flow of shift_register is
    signal q1, q2, q3: std_logic;
begin

end architecture;
```

In our circuit, we have 4 flip-flops. Considering the connections between flip-flops, we can declare the D flip-flops using **port map** function, and make the connections between flip-flop inputs and outputs as in PR 7.9.

PR 7.9 Program 7.9

```
library ieee;
use ieee.std_logic_1164.all;
use work. my_comp_package.all;

entity shift_register is
    port( si, clk: in std_logic;
            so: out std_logic );
end entity;

architecture logic_flow of shift_register is
    signal q1, q2, q3: std_logic;
begin
    dff1: dff port map(si,clk,q1);
    dff2: dff port map(q1,clk,q2);
    dff3: dff port map(q2,clk,q3);
    dff4: dff port map(q3,clk,so);
end architecture;
```

Second Method: In the second method, we do not use a **package** unit. Instead, we make component declarations at the declarative part of the architecture, and the rest is the same as in the first method. The alternative implementation is depicted in PR 7.10.

PR 7.10 Program 7.10

```
library ieee;
use ieee.std_logic_1164.all;
entity shift_register is
    port( si, clk: in std_logic;
            so: out std_logic );
end entity;

architecture logic_flow of shift_register is
    signal q1, q2, q3: std_logic;
    component dff is
        port( d, clk: in std_logic;
                q: out std_logic );
    end component;
begin
    dff1: dff port map(si,clk,q1);
    dff2: dff port map(q1,clk,q2);
    dff3: dff port map(q2,clk,q3);
    dff4: dff port map(q3,clk,so);
end architecture;
```

Example 7.4 Implement the digital circuit of Fig. 7.2 in VHDL. The circuit contains two 4-bit shift registers, one multiplexer, and one 1-bit counter. Use components is your implementation.

Fig. 7.2 Digital circuit for
Example 7.4

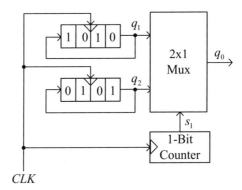

Solution 7.4 First, we separately implement each unique unit of Fig. 7.2 in VHDL. The implementation of the loop-backed shift register in VHDL is given in PR 7.11.

shift_register.vhd

```
library ieee;
use ieee.std_logic_1164.all;

entity shift_register is
     port(clk: in std_logic;
            q: out std_logic );
end entity;

architecture logic_flow of shift_register  is
  signal shift_reg: std_logic_vector(3 downto 0);
begin
   process(clk)
      begin
         if (clk'event and clk='1') then
              shift_reg<=shift_reg(0) & shift_reg(3 downto
1);
           end if;
      end process;
      q<=shift_reg(0) ;
```

PR 7.11 Program 7.11

The implementation of the multiplexer is shown in PR 7.12.

PR 7.12 Program 7.12   mux_2x1.vhd

```vhdl
library ieee;
use ieee.std_logic_1164.all;

entity mux_2x1 is
    port(i0, i1, sel: in std_logic;
            q: out std_logic );
end entity;

architecture  logic_flow  of  mux_2x1
is
begin
  q<=i0 when sel='0' else
        i1;
```

And finally, the implementation of the counter is depicted in PR 7.13.

PR 7.13 Program 7.13   counter.vhd

```vhdl
library ieee;
use ieee.std_logic_1164.all;

entity counter is
    port(clk: in std_logic;
            q: out natural range 0 to 1);
end entity;

architecture logic_flow of counter is
  signal cntr: natural range 0 to 1;
begin
process(clk)
begin
    if (clk'event and clk='1') then
      if(cntr =1) then
        cntr <=0;
      else
        cntr <= cntr +1;
      end if;
    end if;
end process;
  q<= cntr;
end architecture;
```

Once we have all the implementations of circuit units, we can declare all these units as components in the package unit as shown in PR 7.14.

PR 7.14 Program 7.14

my_package.vhd

```
package my_package is
    component shift_register is
        port(clk: in std_logic;
             q: out std_logic );
    end component;
    component mux_2x1 is
        port(i0, i1, sel: in std_logic;
             q: out std_logic );
    end component;
    component counter is
        port(clk: in std_logic;
             q: out natural range 0 to 1);
    end component;
end my_package;
```

After writing the package unit, we can write our main program. For this purpose, we first include our package at the header part, write the entity part and declare the internal signals at the declarative part of the architecture unit as in PR 7.15.

main.vhd

```
library ieee;
use ieee.std_logic_1164.all;
use work. my_package.all;

entity logic_circuit is
    port(clk: in std_logic;
         q0: out std_logic );
end entity;

architecture logic_flow of logic_circuit is
    signal sl, q1, q2: std_logic;
begin

end architecture;
```

PR 7.15 Program 7.15

Considering Fig. 7.2, we instantiate two shift registers, one multiplexer, one counter, and relate the input and output of each component to other as in PR 7.16.

```
library ieee;
use ieee.std_logic_1164.all;
use work. my_package.all;

entity logic_circuit is
    port(clk: in std_logic;
          q0: out std_logic );
end entity;

architecture logic_flow of logic_circuit is
  signal sl, q1, q2: std_logic;
begin
  sr1: shift_register port map(clk,q1);
  sr2: shift_register port map(clk,q2);
  mux: mux_2x1  port map(q1, q2, sl, q0);
  cntr: counter port map(clk,sl);
end architecture;
```

PR 7.16 Program 7.16

Solution 2: We can write the main program without using the package. For this purpose, we first write the entity part, and declare the internal signals and components in the declarative part of the architecture unit and instantiate the components as in PR 7.17.

main.vhd

```
library ieee;
use ieee.std_logic_1164.all;

entity logic_circuit is
    port(clk: in std_logic;
         qo: out std_logic );
end entity;

architecture logic_flow of logic_circuit is

    component shift_register is
        port(clk: in std_logic;
             q: out std_logic );
    end component;

    component mux_2x1 is
        port(i0, i1, sel: in std_logic;
             q: out std_logic );
    end component;

    component counter is
        port(clk: in std_logic;
             q: out natural range 0 to 1);
    end component;

    signal sl, q1, q2: std_logic;

begin
    sr1: shift_register port map(clk,q1);
    sr2: shift_register port map(clk,q2);
    mux: mux_2x1 port map(q1, q2, sl, qo);
    cntr: counter port map(clk,sl);
end architecture;
```

PR 7.17 Program 7.17

Exercise: Implement the circuit of Fig. 7.3, where there are three shift registers and one 2 × 1 multiplexer, in VHDL. Use components in your implementation.

Fig. 7.3 Digital circuit for
VHDL implementation

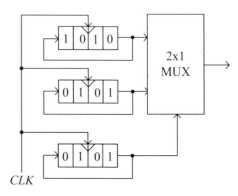

Serial Transfer Unit

Serial transfer units are used to transfer the contents of one shift register to a
second shift register. The logic diagram of the serial transfer unit is depicted in
Fig. 7.4.

Example 7.5 Implement the serial transfer unit shown in Fig. 7.4 in VHDL.
Assume that shift registers contains 8 bits. Use components in your implementation.

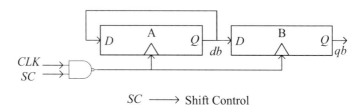

Fig. 7.4 Serial transfer unit

Solution 7.5: First we implement the shift register as in PR 7.18.

shift_register.vhd

```
library ieee;
use ieee.std_logic_1164.all;

entity shift_register is
    port(clk, si: in std_logic;
            so: out std_logic );
end entity;

architecture logic_flow of shift_register is
  signal shift_reg: std_logic_vector(7 downto 0);
begin

  process(clk)
    begin
      if (clk'event and clk='1') then
            shift_reg<= si & shift_reg(7 downto 1);
        end if;
    end process;
      so<=shift_reg(0) ;
end architecture;
```

PR 7.18 Program 7.18

The implementation of the serial transfer unit using components if depicted in PR 7.19.

PR 7.19 Program 7.19

```
                                                    stu.vhdl
library ieee;
use ieee.std_logic_1164.all;

entity logic_circuit is
    port(clk, sc: in std_logic;
            qb: out std_logic );
end entity;

architecture logic_flow of logic_circuit is
    component shift_register is
    port(clk, si: in std_logic;
            so: out std_logic );
    end component;
    signal qa: std_logic;

begin

    sr1: shift_register port map(sc and clk,qa,qa);
    sr2: shift_register port map(sc and clk,qa,qb);

end architecture;
```

Serial Adder Unit

The serial adder is used to sum the contents of two shift registers, and it is depicted in Fig. 7.5.

Example 7.6 Implement the serial adder shown in Fig. 7.5 in VHDL. Use components in your implementation.

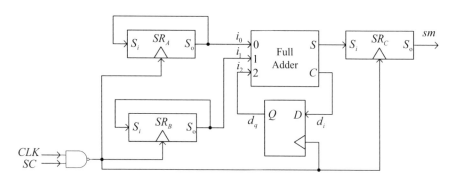

Fig. 7.5 Serial adder unit

Solution 7.6 The shift registers and D flip-flop can be implemented as in PR 7.18 and PR 7.6. The implementation of the full adder is given in PR 7.20.

PR 7.20 Program 7.20

full_adder.vhdl

```
library ieee;
use ieee.std_logic_1164.all;

entity full_adder is
    port(i0, i1, i2: in std_logic;
        s, c: out std_logic );
end entity;

architecture logic_flow of full_adder is
begin
    s<=i0 xor i1 xor i2;
    c<= (i0 and i1) or (i0 and i2) or (i1 and i2);
end architecture;
```

Once we have the implementations of shift register, D flip-flop and full adder circuits, we can implement the circuit of Fig. 7.5 as in PR 7.21.

```
library ieee;
use ieee.std_logic_1164.all;

entity logic_circuit is
    port(clk, sc: in std_logic;
         sm: out std_logic );
end entity;
```

```
architecture logic_flow of logic_circuit is
    signal dq: std_logic;
    signal i0, i1, i2: std_logic; -- Inputs of full adder

    signal s, c: std_logic; -- Outputs of full adder
    signal sm: std_logic; -- Circuit output
    signal qa, qb: std_logic; -- Outputs of SRA and SRB

    component full_adder is
        port(i0, i1, i2: in std_logic;
             s, c: out std_logic );
    end component;

    component shift_register is
        port(clk, si: in std_logic;
             so: out std_logic );
    end component;

    component dff is
        port( d, clk: in std_logic;
              q: out std_logic );
    end component;

begin

    srA: shift_register port map(sc and clk,qa,qa);
    srB: shift_register port map(sc and clk,qb,qb);
    fullAdder: full_adder port map(qa, qb, dq, s, c);
    d_flip_flop: dff(c,dq)
    srC: shift_register port map(sc and clk,s,sm);

end architecture;
```

PR 7.21 Program 7.21

7.3   Functions and Procedures

Functions and procedures are used to write sequential program segments. The functions and procedures are written for the implementation of frequently used and portable algorithms or circuits. Functions and procedures are similar to that of the processes. However, the processes are usually employed inside the architecture unit, on the other hand, functions and procedures are usually placed into the package unit for general use.

7.3.1   Functions

Functions can be placed into the package unit, or they can be written in the declarative part of the architecture unit. If functions are to be placed into the package unit, then a declarative part and a body part of the function should be written and separately placed into the declarative and body part of the package unit.

The syntax of the function inside a packet is shown in PR 7.22.

```
package my_package is

function function_name (input_parameter_list) return data_type;

end my_package;
----------------------------------------------------------------------
package body my_package is

function function_name (input_parameter_list) return data_type is
       variable, constant declarations
begin
       statements
return return_data;
end function_name;

end my_package;
```

PR 7.22 Program 7.22

Example 7.7 The parity calculator function is depicted in PR 7.23. Let's put this function into a package.

```
function parity_calculate(data_vector: std_logic_vector) return std_logic is
  variable parity:std_logic;
begin
    parity:=data_vector(0);
    for index in data_vector'low+1 to data_vector'high loop
      parity:= parity xor data_vector(index);
    end loop;
 return parity;
 end parity_calculate;
```

PR 7.23 Program 7.23

To put the function in PR 7.23 into a package, first we write the header of the function in the declarative part of the package as in PR 7.24.

```
package my_package is

function parity_calculate(data_vector: std_logic_vector) return std_logic;

end my_package;
```

PR 7.24 Program 7.24

In the second step, we place the implementation of the function into the body of the package unit as in PR 7.25.

```
package body my_package is

function parity_calculate(data_vector: std_logic_vector) return std_logic is
    variable parity:std_logic;
begin
    parity:=data_vector(0);
    for index in data_vector'low+1 to data_vector'high loop
      parity:= parity xor data_vector(index);
    end loop;
 return parity;
 end parity_calculate;

end my_package;
```

PR 7.25 Program 7.25

Putting the package header and package body into the same file, we obtain the package unit as in PR 7.26.

```
package my_package is

function parity_calculate(data_vector: std_logic_vector) return std_logic;

end my_package;
------------------------------------------------------------------------
package body my_package is

function parity_calculate(data_vector: std_logic_vector) return std_logic is
    variable parity:std_logic;
begin
    parity:=data_vector(0);
    for index in data_vector'low+1 to data_vector'high loop
       parity:= parity xor data_vector(index);
    end loop;
 return parity;
end parity_calculate;

end my_package;
```

PR 7.26 Program 7.26

Function Call

Once a function is written, it can be called in the architecture body, just typing its name and providing the input values. The return value can be assigned to a signal.

Example 7.8 a_logic_vector <= **conv_std_logic_vector**(an_integer,8);

Example 7.9 The parity calculator function given in PR 7.26 can also be written in the declarative part of the architecture eliminating the use of package at the header of the program. The parity calculator function placed into the architecture part of a VHDL program is depicted in PR 7.27 where it is seen that only function body is placed into the declarative part of the architecture unit. No header declaration of the function as in the packet unit is needed.

```
library ieee;
use ieee.std_logic_1164.all;

entity logic_circuit is
    port( data_vector: in std_logic_vector(15 downto 0);
          prt: out std_logic );
end entity;

architecture logic_flow of logic_circuit is
  function parity_calculate(data_vector: std_logic_vector) return std_logic is
        variable parity:std_logic;
    begin
      parity:=data_vector(0);
      for index in data_vector'low+1 to data_vector'high loop
        parity:= parity xor data_vector(index);
      end loop;
    return parity;
    end parity_calculate;
begin
   prt<=parity_calculate(data_vector);
end architecture;
```

PR 7.27 Program 7.27

While writing a function, we have to pay attention the following items:

(1) Variables are not allowed in the parameter list of the functions, only signals and constants can be used in the parameter list of the functions. Parameter list can be empty as well. Default parameter type is constant.
(2) Return data types should be one of the synthesizable data types available in VHDL.
(3) Range specification is allowed in parameter list, but it is not allowed in the "return data_type" statement at the header of the function.
(4) There is only one return value.
(5) Functions can be nested.

Example 7.9 We can write a function which finds the sum of two logic vectors as in PR 7.28.

```
function sum_logic_vectors(x,y :std_logic_vector(7 downto 0); carry: std_logic)
return std_logic_vector is

    variable cout: std_logic;
    variable cin: std_logic:=carry;
    variable sum: std_logic_vector(8 downto 0);
begin
    sumLoop: for index in 0 to 7 loop
    sum(index):=x(index) xor y(index) xor cin;
    cout:= (x(index) and y(index)) or (x(index) and cin) or (y(index) and cin);
    cin:=cout;
end sumLoop;
    sum(8)=cout;
return sum;
end sum_logic_vectors;
```

PR 7.28 Program 7.28

To use this function, we can place it into a package as in PR 7.29.

```
package my_package is
function sum_logic_vectors(x,y :std_logic_vector(7 downto 0); carry: std_logic)
return std_logic_vector;
end my_package;

package body my_package is
function sum_logic_vectors(x,y :std_logic_vector(7 downto 0); carry: std_logic)
return std_logic_vector is

    variable cout: std_logic;
    variable cin: std_logic:=carry;
    variable sum: std_logic_vector(8 downto 0);
begin
    sumLoop: for index in 0 to 7 loop
    sum(index):=x(index) xor y(index) xor cin;
    cout:= (x(index) and y(index)) or (x(index) and cin) or (y(index) and cin);
    cin:=cout;
end sumLoop;
    sum(8)=cout;
return sum;
end sum_logic_vectors;
end my_package;
```

PR 7.29 Program 7.29

The package unit can be written in a separate file or we can use it in the main program before the **entity** part. If it is used in the main program, then we need to write

use work.my_package.**all**

after declarations of libraries in the main program. In PR 7.30. the package unit is included in the main program, and the function is called in the body part of the architecture unit.

```
package my_package is
function sum_logic_vectors(x,y :std_logic_vector(7 downto 0); carry: std_logic)
return std_logic_vector;
end my_package;

package body my_package is
function sum_logic_vectors(x,y :std_logic_vector(7 downto 0); carry: std_logic)
return std_logic_vector is

    variable cout: std_logic;
    variable cin: std_logic:=carry;
    variable sum: std_logic_vector(8 downto 0);
begin
    sumLoop: for index in 0 to 7 loop
    sum(index):=x(index) xor y(index) xor cin;
    cout:= (x(index) and y(index)) or (x(index) and cin) or (y(index) and cin);
    cin:=cout;
end sumLoop;
    sum(8)=cout;
return sum;
end sum_logic_vectors;
end my_package;

-- Package declarations must be made before the inclusion of other library packages

library ieee;
use ieee.std_logic_1164.all;

use work.my_package.all; -- Necessary to use the package

entity logic_circuit is
    port( d1, d2: in std_logic_vector(7 downto 0);
            q: out std_logic_vector(8 downto 0) );
end entity;

architecture logic_flow of logic_circuit is
begin
    q<= sum_logic_vectors(d1,d2);
end architecture;
```

PR 7.30 Program 7.30

7.3.2 Operator Overloading

To increase the argument variability of the operators, we use the operator overloading approach. For this purpose, we write a function and the name of the function is chosen as "operator" where operator can be $+, -, *$, and $/$ etc. The syntax of the operator overloading is depicted in PR 7.31.

```
function "operator" (parameters) return data_type is
   declarations
begin
   sequential statements
end function;
```

PR 7.31 Program 7.31

Example 7.10 The summation operator "+" is overloaded for two logic vectors as in PR 7.32.

```
library ieee;
use ieee.std_logic_1164.all;

entity logic_circuit is
    port( d1,d2: in std_logic_vector(15 downto 0);
            sm: out std_logic_vector(16 downto 0) );
end entity;

architecture logic_flow of logic_circuit is
    function "+" (x,y :std_logic_vector(16 downto 0); carry: std_logic) return
std_logic_vector is

    variable cout: std_logic;
    variable cin: std_logic:=carry;
    variable sum: std_logic_vector(8 downto 0);
begin
    sumLoop: for index in 0 to 7 loop
    sum(index):=x(index) xor y(index) xor cin;
    cout:= (x(index) and y(index)) or (x(index) and cin) or (y(index) and cin);
    cin:=cout;
    end sumLoop;
        sum(8)=cout;
    return sum;
end sum_logic_vectors;


begin
    sm<= d1+d2;
end architecture:
```

PR 7.32 Program 7.32

Example 7.11 In PR 7.33, operator overloading is illustrated for "+" considering different operand data types. However, we did not implement the function bodies inside the package body. Since the aim of the example is to illustrate the concept of overloading rather than implementation.

```
library ieee;
use ieee.std_logic_1164.all;

package my_arithmetic is
  function "+" (L: std_logic_vector; R: std_logic_vector) return integer;
  function "+" (L: std_logic_vector; R: std_logic_vector) return std_logic_vector;
  function "+" (L: std_logic_vector; R: integer) return std_logic_vector;
  function "+" (L: integer; R: std_logic_vector) return std_logic_vector;
end my_arithmetic;

package body my_arithmetic is
  -- Here we write the implementations of the functions declared in the declarative part of
    the package
end my_arithmetic;

use work. my_arithmetic.all;

entity example is
  port (x_vec, y_vec: in  std_logic_vector (7 downto 0);
        wo_bus , yo_bus, zo_bus: out std_logic_vector (7 downto 0);
        x_int, y_int: in integer;
        zo_int: out integer);
end example;

architecture of example is
  begin
      zo_int <= x_vec + y_vec;
      zo_bus <= x_vec + y_vec;
      yo_bus <= x_vec + y_int;
      wo_bus <= x_int + y_vec;
end architecture;
```

PR 7.33   Program 7.33

7.4   Procedures

Procedures are program units similar to the functions and they are used for writing sequential program segments. Functions return a single parameter, on the other hand, procedures can return more than one parameter. In addition, in the parameter list of the procedures, we can use variables, signals and constants.

The procedures can be written in the declarative part of the architecture unit, or they can be declared in the declarative part of the package, and their body can be written in the body part of the package unit. The syntax of the procedure unit is shown in PR 7.34.

```
procedure procedure_name(input and output parameters) is
    declarations
begin
    sequential statements
end procedure;
```

PR 7.34 Program 7.34

Example 7.12 The procedure, placed into a package, that inputs three integers and finds the minimum and maximum of these three integers is depicted in PR 7.35.

```
package my_package is
    procedure min_max(signal w, x, y :in integer; signal min, max: out integer);
end my_package;

package body my_package is
procedure min_max(variable w, x, y :in integer; variable min, max: out integer) is
    variable temp1, temp2: integer;
begin
    temp1:= w;
    temp2:= w;
    if (x > temp1) then
        temp1 := x;
    end if;
    if (y> temp1) then
        temp1 := y;
    end if;
    if (x< temp2) then
        temp2 := x;
    end if;
    if (y< temp2) then
        temp2 := y;
    end if;
    max:=temp1;
    min:=temp2;
    end procedure;
end my_package;
```

PR 7.35 Program 7.35

Including the package at the header of the main program, we can call the procedure in the body of the architecture unit as in PR 7.36.

```
library ieee;
use ieee.std_logic_1164.all;
use work.my_package.all; -- Necessary to use the package

entity logic_circuit is
    port( w, x, y: in integer;
            min, max: out integer );
end entity;

architecture logic_flow of logic_circuit is
begin
    min_max(w,x,y,min,max);
end architecture;
```

PR 7.36   Program 7.36

7.4.1   Differences Between a Function and a Procedure

(1) A function can return only a single parameter, and input parameter list of a function can contain signals and constants only, variables are not allowed in the parameter list of a function.
(2) A procedure can accept any number of inputs and can return any number of parameters. The input parameter list of the procedure can contain variables, signals and constants.
(3) When a function is called, its return value should be assigned to a parameter.
(4) Both function and procedures can be placed inside a package, or they can be written in the declarative part of the architecture.

Problems

(1) Implement the 4-bit shift register shown in Fig. 7.P1 in VHDL. Use components in your implementation.

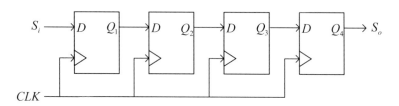

Fig. 7.P1   4-bit shift register

(2) Implement the circuit which contains 8-bit shift registers shown in Fig. 7.P2 in VHDL. Use components in your implementation.

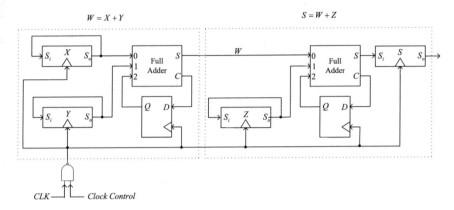

Fig. 7.P2 Digital circuit for VHDL implementation

(3) Implement the circuit shown in Fig. 7.P3 where shift registers contain 8-bits in VHDL. Use components in your implementation.

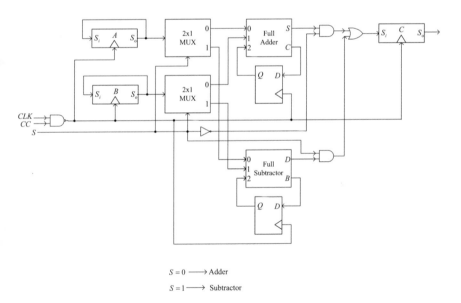

Fig. 7.P3 Digital circuit for VHDL implementation

(4) Implement the logic module shown in Fig. 7.P4 in VHDL. Use components in your implementation.

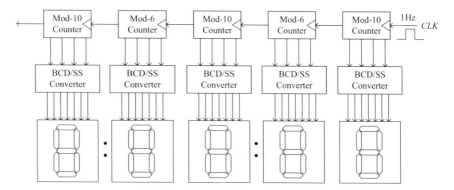

Fig. 7.P4 Digital circuit for VHDL implementation

(5) Write a function that finds the XOR of two 8-bit std_logic_vectors. Put the function into a package and call your function in your main program.

(6) Write a procedure that sorts three integers in ascending order. Implement your procedure directly in the declarative part of the architecture unit.

(7) Can I use a variable in the parameter list of a function?

(8) Can I use a variable in the parameter list of a procedure?

Chapter 8
Fixed and Floating Point Numbers

In this chapter, fixed and floating point number implementations in VHDL will be explained. Fixed and floating number formats enable the VHDL programmer to implement the fractional numbers in an easy manner. Fixed point number format is used to implement the fractional integers; on the other hand, floating point format is used to implement the real numbers in FPGA platform.

8.1 Fixed Point Numbers

Fixed point number format is introduced in VHDL-2008. For this reason, during the project creation, VHDL-2008 language should be chosen for editing. Some of the VHDL development platforms may not support fixed point number formats. Fixed point number format is supported in ISE XILINX platform.

Fixed point numbers can be signed or unsigned. A signed number in base-2 can be expressed in general form

$$(\cdots d_2 d_1 d_0 \cdot f_{-1} f_{-2} f_{-3} \cdots)_2$$

which can be converted to decimal number as

$$\cdots + d_2 2^2 + d_1 2^1 + d_0 2^0 + f_{-1} 2^{-1} + f_{-2} 2^{-2} + f_{-3} 2^{-3} + \cdots$$

To define the signed fixed and unsigned fixed point numbers, we use the syntax

signal/**variable** snum : **sfixed** (a **downto** b);

signal/**variable** unum : **ufixed**(a **downto** b);

© Springer Nature Singapore Pte Ltd. 2019
O. Gazi, *A Tutorial Introduction to VHDL Programming*,
https://doi.org/10.1007/978-981-13-2309-6_8

where a + 1 and |b| indicates the number of bits used for the integer and fractional part of the numbers if a > 0 and b < 0. Signed numbers are represented in 2's complement form.

To be able to use the fixed point number format in our VHDL code, we need to include the lines

library ieee_proposed;

use ieee_proposed.fixed_pkg.all;

at the header of the VHDL program.

Example 8.1 In the VHDL statement

$$\textbf{signal } \text{num} : \textbf{ufixed } (4 \textbf{ downto } - 3)$$

the parameter 'num' represents an 8-bit number in the form

$$d_4 d_3 d_2 d_1 d_0 . f_{-1} f_{-2} f_{-3}$$

where the first 5-bit is used for the integer part, and the 3-bit after point is used for the fractional part.

Example 8.2 Considering the VHDL statement

$$\textbf{signal } \text{num} : \textbf{ufixed}(4 \textbf{ downto } - 3) := \text{"10101101"}$$

find the decimal equivalent of "num".

Solution 8.2 The binary string

$$10101101$$

can be converted to decimal as

$$\underbrace{10101}_{\substack{integer \\ part}} . \underbrace{101}_{\substack{fractional \\ part}} \rightarrow \underbrace{1 \times 2^4 + 0 \times 2^3 + 1 \times 2^2 + 0 \times 2^1 + 1 \times 2^0}_{integer\,part}$$

$$+ \underbrace{1 \times 2^{-1} + 0 \times 2^{-2} + 1 \times 2^{-3}}_{fractional\,part} \rightarrow \underbrace{21}_{\substack{integer \\ part}} + \underbrace{0.625}_{\substack{fractional \\ part}} \rightarrow 21.625$$

Example 8.3 Considering the VHDL statement

$$\textbf{signal } \text{num} : \textbf{sfixed}(4 \textbf{ downto } - 3) := \text{"10101101"}$$

find the decimal equivalent of "num".

Solution 8.3 The binary string

$$10101101$$

represent a negative number in 2's complement form. To find the negative number in decimal form, we first take the 2's complement of

$$10101101$$

resulting in

$$01010011$$

which can be converted to decimal as

$$\underbrace{01010}_{\substack{integer \\ part}} . \underbrace{011}_{\substack{fractional \\ part}} \rightarrow \underbrace{0 \times 2^4 + 1 \times 2^3 + 0 \times 2^2 + 1 \times 2^1 + 0 \times 2^0}_{integer\, part}$$

$$+ \underbrace{0 \times 2^{-1} + 1 \times 2^{-2} + 1 \times 2^{-3}}_{fractional\, part} \rightarrow \underbrace{10}_{\substack{integer \\ part}} + \underbrace{0.375}_{\substack{fractional \\ part}} \rightarrow 10.375$$

Thus, the binary string

$$10101101$$

represent the negative number

$$-10.375.$$

Example 8.4 In the VHDL statement

$$\textbf{signal}\,\text{num} : \textbf{ufixed}(-3\,\textbf{downto} - 5)$$

the parameter 'num' represents a positive number in the form

$$0.00f_{-3}f_{-4}f_{-5}$$

whose decimal equivalent can be calculated as

$$f_{-3} \times 2^{-3} + f_{-4} \times 2^{-4} + f_{-5} \times 2^{-5}.$$

Example 8.5 In the VHDL statement

$$\textbf{signal}\,\text{num} : \textbf{ufixed}(-3\,\textbf{downto} - 5) := \text{``101''}$$

the parameter 'num' represents a positive number in the form

$$0.00101$$

whose decimal equivalent can found as

$$0.00101 \rightarrow 1 \times 2^{-3} + 1 \times 2^{-5} \rightarrow 0.15625.$$

Example 8.6 In the VHDL statement

$$\textbf{signal}\, \text{num} : \textbf{sfixed}(-3\,\textbf{downto} - 5) := \text{"111011"}$$

the parameter 'num' represents a negative number in the form

$$1.11011$$

whose decimal equivalent can be found as

$$1.11011 \xrightarrow{\overset{2's}{complement}} 0.00101 \xrightarrow{Decimal} 0.15625 \rightarrow -0.15625.$$

Example 8.7 In the VHDL statement

$$\textbf{signal}\, \text{num} : \textbf{sfixed}(-3\,\textbf{downto} - 5) := \text{"011"}$$

the parameter 'num' represents number in the form

$$0.00011$$

whose decimal equivalent can be found as

$$0.00011 \rightarrow 0.09375.$$

Example 8.8 In the VHDL statement

$$\textbf{signal}\, \text{num} : \textbf{sfixed}(3\,\textbf{downto} - 1) := \text{"11100"}$$

the parameter 'num' represents a negative number in the form

$$1110.0$$

whose decimal equivalent can be found as

$$1110.0 \xrightarrow{\overset{2's}{complement}} 00010.0 \xrightarrow{Decimal} 2 \rightarrow -2.$$

8.1.1 Type Conversion Functions

For fixed point number, the type conversion functions

$$\textbf{to\_ufixed}(\text{number}, a, -b) \quad \textbf{to\_sfixed}(\text{number}, a, -b) \quad a > 0, \ b > 0$$

are available, and these functions are used to convert a fractional number to a binary string which is interpreted according to fixed point number format.

Example 8.9 In PR 8.1, the fractional number 12.25 is represented using fixed point format via two different approach. In the first approach, 12.25 is directly converted to fixed point format. In the second approach, we first convert 12.25 to a binary string, i.e., "01100_010", and then convert it to fixed point format.

PR 8.1 Program 8.1

```
signal num1: ufixed (4 downto -3);
num1<= to_ufixed(12.25, 4, -3);

signal num2: ufixed (4 downto -3);
num2<= to_ufixed("01100_010", 4, -3);
```

Fixed point numbers can be converted to other data formats. For this purpose, we use the conversion functions

to_slv() → The binary string representing a **signed** or an **unsigned fixed point** number is interpreted as a **std_logic_vector**.

to_signed() → The binary string representing a **signed** or an **unsigned fixed point** number is interpreted as a binary string representing a signed integer.

to_unsigned() → The binary string representing a **signed** or an **unsigned fixed point** number is interpreted as a binary string representing an unsigned integer.

to_integer() → The binary string representing a **signed** or an **unsigned fixed point** number is interpreted as a binary string representing an integer.

to_std_logic_vector() → Same as **to_slv()**.

add_sign() → An unsigned fixed point number is converted into a signed fixed point number. To achieve this, 2's complement of the positive number is taken so that the result represents a negative number.

Example 8.10 In PR 8.2, a fixed point number represented by the binary string "11000111" is converted into numbers in data types **std_logic_vector**, **signed**, **integer**. Considering PR 8.2, find the binary strings for num2, num3, and num4.

PR 8.2 Program 8.2

> **signal** num1: **ufixed** (4 **downto** -3):="11000111";
> **signal** num2: **std_logic_vector**(7 **downto** 0);
> **signal** num3: **signed**(7 **downto** 0);
> **signal** num4: **integer range** -128 to 127;
>
> num2<= **to_slv**(num1);
> num3<= **to_signed**(num1);
> num4<= **to_integer**(num1);

Solution 8.10 The conversion functions given in PR 8.2 does not alter the binary string, just the interpretation of the string changes. So for this reason, the binary strings for num2, num3, and num3 are the same as num1, i.e., "11000111".

Example 8.11 In PR 8.3, an unsigned fixed point number represented by the binary string "01000111" is converted into a signed fixed point number using the conversion function **add_sign()**. Considering PR 8.3, find the binary strings for num2.

PR 8.3 Program 8.3

> **signal** num1: **ufixed** (4 **downto** -3):="11000111";
> **signal** num2: **sfixed** (5 **downto** -3);
>
> num2<= **add_sign**(num1);

Solution 8.11 In PR 8.3, num1 equals to 24.875. On the other hand, num2 equals to −24.875. We can find the binary string for num2 as

$$11000111 \rightarrow 2's \text{ complement} \rightarrow 00111001 \rightarrow \text{prefix } '1' \text{ to the front of string}$$
$$\rightarrow 100111001$$

8.1.2 Operators for Fixed Point Numbers

Since fixed point numbers are represented by binary strings, then logical operations can be performed between fixed point number. The logical operators that can be used for fixed point numbers are:

and, nand, or, nor, xor, xnor, not

Arithmetic operators can also be used between fixed point numbers. The arithmetic operators that can be used between fixed point numbers are

$+, \quad -, \quad *, \quad /, \quad$ **rem, mod, abs**

The comparison operators that can be used between fixed point numbers are:

$$=, \quad >=, \quad <=, \quad <, \quad >, \quad /=$$

The shift functions that can operate on fixed point numbers are

sll, srl, rol, ror, sla, sra

8.1.3 Arithmetic Operations with Fixed Point Numbers and Sizing Rules

Let num1 and num2 be two unsigned fixed point numbers defined as

$$\textbf{signal } num1 : (s/u) \textbf{ fixed}(a \textbf{ downto } b); \quad b < 0$$
$$\textbf{signal } num2 : (s/u) \textbf{ fixed}(c \textbf{ downto } d); \quad d < 0$$

When arithmetic operations are performed using num1 and num2, the result is a fixed point defined as

$$\textbf{signal } res : (s/u) \textbf{ fixed } (e \textbf{ downto } f) \quad f < 0.$$

To prevent the loss of data, the range values e and f of the res should be chosen properly. The rule to decide on the values of e and f is tabulated in Table 8.1

Table 8.1 Range calculation for fixed point arithmetic operations

Operation	Range of result	
num1 + num2	e = max(a,c) + 1	f = min(b,d)
num1 − num2	e = max(a,c) + 1	f = min(b,d)
num1 * num2	e = a+c + 1	f = b+d
num1 **rem** num2	e = min(a,c)	f = min(b,d)
num1 **mod** num2 (signed)	e = min(a,c)	f = min(b,d)
num1 **mod** num2 (unsigned)	e = c	f = min(b,d)
num1/num2 (signed)	e = a − d + 1	f = b − c
num1/num2 (unsigned)	e = a − d	f = b − c − 1
−num1	e = a + 1	f = b
abs(num1)	e = a + 1	f = b
reciprocal(num1) (signed)	e = −b + 1	f = −a
reciprocal(num1) (unsigned)	e = −b	f = −a − 1

Example 8.12 For the two fixed point numbers

$$\textbf{signal}\, x : \textbf{ufixed}\, (4\, \textbf{downto} - 3);$$
$$\textbf{signal}\, y : \textbf{ufixed}\, (8\, \textbf{downto} - 7);$$

the result of x*y is a number with data type

$$\textbf{ufixed}\, (4 + 8 + 1\, \textbf{downto} - 3 - 7) \rightarrow \textbf{ufixed}\, (13\, \textbf{downto} - 10)$$

and the result of x + y, or x − y is a number with data type

$$\textbf{ufixed}(8 + 1\, \textbf{downto} - 7) \rightarrow \textbf{ufixed}(9\, \textbf{downto} - 7)$$

Similarly, the result of x/y is a number with data type

$$\textbf{ufixed}(4 - (-7)\, \textbf{downto} - 3 - 8 - 1) \rightarrow \textbf{ufixed}(11\, \textbf{downto} - 12)$$

8.1.4   Automatic Resizing

If you multiply or divide any two fixed point numbers, either you need to calculate the size of the result, or, you need to use a large size for the result to prevent the overflow and loss of information. The size of the result can be estimated using the special functions

ufixed_high,   ufixed_low,   sfixed_high,   sfixed_low

Let's say that we want to multiply and divide two fixed point numbers, and assign results to fixed point parameters. The size of the result can be calculated as illustrated in PR 8.4.

```
library ieee_proposed;
use ieee_proposed.fixed_pkg.all;

entity fixedTest is
end fixedTest;

architecture Behavioral of fixedTest is
 signal num1: sfixed(8 downto -4);
 signal num2: sfixed(15 downto -6);

 signal result1: sfixed(sfixed_high(num1'high,num1'low,'/',num2'high,num2'low)
            downto sfixed_low(num1'high,num1'low,'/',num2'high,num2'low));

 signal result2: sfixed(sfixed_high(num1'high,num1'low,'*',num2'high,num2'low)
            downto sfixed_low(num1'high,num1'low,'*',num2'high,num2'low));
begin
 num1<=to_sfixed(12.25,8,-4);
 num2<=to_sfixed(4.25,15,-6);

 result1<=num1/num2;
 result2<=num1*num2;

end Behavioral;
```

PR 8.4   Program 8.4

Example 8.13 Write a VHDL program that takes two signed fixed point numbers from input ports and calculates the multiplication and division results of these two numbers and sends the results to the output ports.

Solution 8.13 The requested program can be written as in PR 8.5.

```
library ieee_proposed;
use ieee_proposed.fixed_pkg.all;

package my_package is
    constant num1_high: integer:=8;
    constant num1_low: integer:=-6;
    constant num2_high: integer:=12;
    constant num2_low: integer:=-3;
    subtype my_sfixed is
      sfixed(sfixed_high(num1_high,num1_low,'/',num2_high,num2_low)
        downto
      sfixed_low(num1_high,num1_low,'/',num2_high,num2_low));
end my_package;

package body my_package is
end my_package;

library ieee_proposed;
use ieee_proposed.fixed_pkg.all;
use work.my_package.all;

entity fixedTest is
port (num1: in sfixed(num1_high downto num1_low);
      num2: in sfixed(num2_high downto num2_low);
      result1: out my_sfixed;
      result2: out my_sfixed);
    end fixedTest;

architecture Behavioral of fixedTest is

begin

  result1<=num1/num2;
  result2<=num1*num2;

end Behavioral;
```

PR 8.5 Program 8.5

8.1.5 *Resize Function*

We know that performing arithmetic operations on fixed point numbers result in another fixed point number with larger size. If we are sure that overflow does now occur after performing arithmetic operations on fixed point numbers, we can make the size of the resulting fixed point number the same as the fixed point numbers involved in operations.

For instance, assume that two 8-bit unsigned fixed point numbers are to be summed, and these numbers can be at most 100, then the sum of these numbers can be at most 200 which can be represented using 8-bit.

The declarative statement for the **resize** function is as

function resize(argument, left size, right size, round style, overflow style)

where round style is a Boolean variable, and when round style is true then fixed round is adapted for argument, otherwise argument is truncated in case of overflow.

In addition, if overflow style is true, then fixed saturate mode is chosen, which implies that the maximum possible number is returned, if the argument is too large to represent via the given size values. Otherwise, fixed wrap mode is adapted, which implies that the top bits are truncated in case of overflow.

Example 8.14 In PR 8.6, we create a number vector consisting of fixed point number, and find the sum of numbers and send the summation result to an output port.

```
library ieee_proposed;
use ieee_proposed.fixed_pkg.all;

entity fixedTest is
  port(sum: out sfixed(12,-3) );
end fixedTest;

architecture Behavioral of fixedTest is

type sfixed_vector is array (natural range <>) of sfixed(12 downto -3);

signal data_vector: sfixed_vector(0 to 15);
signal sum_result: sfixed_vector(0 to 15);

begin

data_vector<=(
to_sfixed(12.25,12,-3), to_sfixed(8,12,-3),    to_sfixed(-6.5,12,-3),  to_sfixed(3.5,12,-3),
to_sfixed(2.25,12,-3),  to_sfixed(8,12,-3),    to_sfixed(16,12,-3),    to_sfixed(23,12,-3),
to_sfixed(7,12,-3),     to_sfixed(-4.5,12,-3), to_sfixed(-18,12,-3),   to_sfixed(3.1,12,-3),
to_sfixed(23.25,12,-3), to_sfixed(-5,12,-3),   to_sfixed(-6.5,12,-3),  to_sfixed(5.5,12,-3)
);
sum_result(0)<=data_vector(0);

fixedSum: for index in 1 to 15 generate
  sum_result(index)<=resize(sum_result(index-1)+data_vector(index), 12, -3);
end generate fixedSum;

sum<=sum_result(15);

end Behavioral;
```

PR 8.6 Program 8.6

8.1.6   Add Cary Procedure

In **ieee_proposed** library, there are some built-in functions and procedures for fixed point operations. One of the procedure is the add-carry procedure whose declarative statement is

$$\textbf{add\_carry}(\text{inp1, inp2, cin, out1, cout})$$

where out1 = inp1 + inp2 + cin and although inputs can be any objects, such as signal, constant, variable, the outputs out1 and cout are variable objects. The complete declarative part of the add_carry procedure have two different forms as shown below

> **procedure** add_carry(
>   L, R : **in unresolved_ufixed**;
>   c_in : **in std_ulogic**;
>   result : **out unresolved_ufixed**;
>   c_out : **out std_ulogic**);
>
> **procedure** add_carry(
>   L, R : **in unresolved_sfixed**;
>   c_in : **in std_ulogic**;
>   result : **out unresolved_sfixed**;
>   c_out : **out std_ulogic**);

In **ieee_proposed** library, the data types "**unresolved_u/sfixed**" and "**s/ufixed**" are defined as

> **type** unresolved_ufixed **is array**(**integer range** < >)of **std_ulogic**;
> **type** unresolved_sfixed **is array**(**integer range** < >)of **std_ulogic**;
>
> **subtype** ufixed is **unresolved_ufixed**;
> **subtype** sfixed is **unresolved_sfixed**;

Example 8.15 In PR 8.7, the use of add_carry() procedure is explained with an example.

```
library ieee_proposed; library ieee; use ieee_proposed.fixed_pkg.all;

package my_package is
  type ufixed_vector is array (natural range <>) of ufixed(12 downto -3);
end my_package;

library ieee_proposed; library ieee;   use ieee_proposed.fixed_pkg.all;
use ieee.std_logic_1164.all;

use work.my_package.all;

entity fixedTest is
     port(sum: out ufixed(12,-3) );
end fixedTest;

architecture Behavioral of fixedTest is

    signal data_array: ufixed_vector (0 to 15);

  function fixed_point_sum(data_vector: ufixed_vector) return ufixed is

    variable result: ufixed(12,-3);
    variable c_out: std_ulogic;
    variable c_in: std_ulogic;
    variable sum_result: ufixed_vector(0 to 15);
begin

c_in:='0';
sum_result(0):=to_ufixed(0,12,-3);

for index in 1 to 15 loop
  add_carry(sum_result(index-1),data_vector(index),c_in,sum_result(index),c_out);
  c_in:=c_out;
end loop;

    result:=sum_result(15);
    return result;
end fixed_point_sum;

  begin
  data_array<=(
  to_ufixed(12.25,12,-3), to_ufixed(8,12,-3), to_ufixed(6.5,12,-3), to_ufixed(3.5,12,-3),
  to_ufixed(2.25,12,-3), to_ufixed(8,12,-3), to_ufixed(16,12,-3), to_ufixed(23,12,-3),
  to_ufixed(7,12,-3), to_ufixed(4.5,12,-3), to_ufixed(18,12,-3), to_ufixed(3.1,12,-3),
  to_ufixed(23.25,12,-3), to_ufixed(5,12,-3), to_ufixed(6.5,12,-3), to_ufixed(5.5,12,-3)
  );
      sum<=fixed_point_sum(data_array);
end Behavioral;
```

PR 8.7 Program 8.7

8.1.7 Divide Function

The divide function has the declarative statement defined as **function** divide (l, r: **unresolved_ufixed**; **constant** round_style: fixed_round_style_type: = fixed_round_style; **constant** guard_bits: **natural**: = fixed_guard_bits) **return unresolved_ufixed**;

When divide function is used for the division of two ufixed numbers as

$$\mathbf{ufixed}(a \text{ downto } b)/\mathbf{ufixed}(c \text{ downto } d)$$

the result will be an ufixed number as

$$\mathbf{ufixed}(a - d \text{ downto } b - c - 1).$$

The divide function for sfixed point numbers has the declarative part given as **function** divide (l, r: **unresolved_sfixed**; **constant** round_style: fixed_round_style_type: = fixed_round_style; **constant** guard_bits: **natural**: = fixed_guard_bits) **return unresolved_sfixed**;

When divide function is used for the division of two sfixed numbers as

$$\mathbf{sfixed}(a \text{ downto } b)/\mathbf{sfixed}(c \text{ downto } d)$$

the result will be an sfixed number as

$$\mathbf{sfixed}(a - d + 1 \text{ downto } b - c).$$

8.1.8 Reciprocal Function

The reciprocal function returns the inverse of a number, i.e., return $1/x$. The reciprocal function for ufixed numbers has the declarative statement defined as **function** reciprocal (arg: **unresolved_ufixed**; constant round_style: **fixed_round_style_type** := fixed_round_style; **constant** guard_bits: **natural**: = fixed_guard_bits) **return unresolved_ufixed**;

When reciprocal function is used for an ufixed number as

$$1/\mathbf{ufixed}(a \text{ downto } b)$$

the result will be an ufixed number as

$$\mathbf{ufixed}(-b \text{ downto } - a - 1)$$

The reciprocal function for sfixed numbers has the declarative statement defined as

function reciprocal (arg: **unresolved_sfixed**; constant round_style: **fixed_round_style_type**: = fixed_round_style; **constant** guard_bits: **natural**: = fixed_guard_bits) **return unresolved_sfixed**;

When reciprocal function is used for a sfixed numbers as

$$1/\mathbf{sfixed}(a\,\mathbf{downto}\,b)$$

the result will be a sfixed number as

$$\mathbf{sfixed}(-b+1\,\mathbf{downto}-a).$$

8.1.9   Remainder Function

The remainder function is used to find the remainder after division of two fixed point numbes. The declarative statement of the remainder function for ufixed number is as

function remainder(l, r: **unresolved_ufixed**; **constant** round_style: **fixed_round_style_type**: = fixed_round_style; **constant** guard_bits: **natural**: = fixed_guard_bits) **return unresolved_ufixed**;

When remainder function is used for ufixed numbers as

$$\mathbf{ufixed}(a\,\mathbf{downto}\,b)\,\mathbf{rem}\,\mathbf{ufixed}(c\,\mathbf{downto}\,d)$$

the result will be an ufixed number as

$$\mathbf{ufixed}(\mathrm{minimum}(a,c)\,\mathbf{downto}\,\mathrm{minimum}(b,d)).$$

The declarative statement of the remainder function for sfixed numbers is as

function remainder(l, r: **unresolved_sfixed**; **constant** round_style: **fixed_round_style_type**: = fixed_round_style; **constant** guard_bits: **natural**: = fixed_guard_bits) **return unresolved_sfixed**;

When remainder function is used for sfixed numbers as

$$\mathbf{sfixed}(a\,\mathbf{downto}\,b)\,\mathbf{rem}\,\mathbf{sfixed}(c\,\mathbf{downto}\,d)$$

the result will be an sfixed number as

$$\mathbf{sfixed}(\mathrm{minimum}(a,c)\,\mathbf{downto}\,\mathrm{minimum}(b,d)).$$

The implementation of overloaded **rem** operator is given in PR 8.8.

```
function "rem" (l, r : unresolved_ufixed) return unresolved_ufixed is
begin
return remainder (l, r);
end function "rem";

function "rem" (l, r : unresolved_sfixed) return unresolved_sfixed is
begin
return remainder (l, r);
end function "rem";
```

PR 8.8 Program 8.8

Note that an overloaded operator is different than an overloaded function in terms of input accepting. For instance, the overloaded operator in PR 8.8 is used as "arg1 **rem** arg2", on the other hand, the overloaded function remainder is called as "**remainder**(arg1, arg2)".

8.1.10   Scalb Function

The scalb function is used to multiply or divide a fixed point number by a number which is power of two. The function has the different overloaded declarations as outlined below:

function scalb (x : unresolved_ufixed; N : **integer**)**return unresolved_ufixed**;
function scalb (x : unresolved_ufixed; N : **signed**)**return unresolved_ufixed**;
function scalb (x : unresolved_sfixed; N : **integer**)**return unresolved_sfixed**;
function scalb (x : unresolved_sfixed; N : **signed**)**return unresolved_sfixed**;

If x and N are the inputs of the function, then the function output happens to be

$$y = x \times 2^N.$$

The width of the input equals to the width of the output with the binary point moved.

Example 8.16 In PR.9, the use of **scalb** function is illustrated with an example where an unsigned fixed point number of type

$$\textbf{ufixed}(2\,\textbf{downto} - 2)$$

with value "000.10" corresponding to 0.5 is multiplied by 2^4 using the **scalb** function yielding 8.0 which can be represented by "1000.0" with the data type

$$\mathbf{ufixed}(3\,\mathbf{downto}-1).$$

In PR 8.9, we use **resize** function to change the size of the resulting number. If **resize** function is not used, then the value of the binary string may not be accurately computed by the programmer.

PR 8.9 Program 8.9

```
library ieee_proposed;
use ieee_proposed.fixed_pkg.all;

entity fixedTest is
  port( sum: out ufixed(3 downto -1) );
end fixedTest;

architecture Behavioral of fixedTest is
  signal data: ufixed(2 downto -2);
  subtype ufixed2_2 is ufixed(2 downto -2);

begin
  data<=ufixed2_2'("00010");
  sum<=resize(scalb(data,4),3,-1);
end Behavioral;
```

8.1.11 Maximum and Minimum Function

The maximum and minimum functions are used to find the maximum or minimum of two fixed point numbers. The declarative statements of these functions are as

function maximum (l, r: **unresolved_ufixed**) **return unresolved_ufixed**

function maximum (l, r: **unresolved_sfixed**) **return unresolved_sfixed**

function minimum (l, r: **unresolved_ufixed**) **return unresolved_ufixed**

function minimum (l, r: **unresolved_sfixed**) **return unresolved_sfixed**

In PR 8.10, the use of the maximum and minimum functions are explained via an example.

PR 8.10 Program 8.10

```
library ieee_proposed;
use ieee_proposed.fixed_pkg.all;

entity fixedTest is
  port( inp1: in ufixed(4 downto -3);
        inp2: in ufixed(4 downto -3);
        max_num: out ufixed(4 downto -3);
        min_num: out ufixed(4 downto -3));
end fixedTest;

architecture Behavioral of fixedTest is
begin
  max_num<=maximum(inp1, inp2);
  min_num<=minimum(inp1, inp2);
end Behavioral;
```

8.1.12 'Find Left Most' and 'Find Right Most' Functions

The functions **find_leftmost** and **find_rightmost** are used to find the most and least significant bits of a fixed point number. The declarative statement for these functions are as follows:

function find_leftmost (arg: **unresolved_ufixed**; y: **std_ulogic**) **return integer**;
function find_leftmost (arg: **unresolved_sfixed**; y: **std_ulogic**) **return integer**;

function find_rightmost (arg: **unresolved_ufixed**; y: **std_ulogic**) **return integer**;
function find_rightmost (arg: **unresolved_sfixed**; y: **std_ulogic**) **return integer**;

These functions are used to find the location of 'y' in the fixed point number. If 'y' is not found in the number, then **find_leftmost** function returns "arg'low-1", and **find_rightmost** function returns "arg'high + 1". For **find_leftmost** function, the search starts from the most significant bit and decreases, and similarly for **find_rightmost** function, the search starts from the least significant bit and increases.

In PR 8.11, the use of the **find_leftmost** and **find_rightmost** functions are illustrated with an example.

PR 8.11   Program 8.11

```
library ieee_proposed;
use ieee_proposed.fixed_pkg.all;

entity fixedTest is
  port( loc1: out integer;
        loc2: out integer);
end fixedTest;

architecture Behavioral of fixedTest is
  signal data: ufixed(9 downto -6);
  subtype ufixed9_6 is ufixed(9 downto -6);

begin
  data<=ufixed9_6'("0011000000000110");
  loc1<=find_leftmost(data,'1'); -- Returns 7
  loc2<=find_rightmost(data,'1'); -- Returns -5
end Behavioral;
```

8.1.13   ABS Function

The **abs** function is used to change the sign of negative numbers. The declarative statement of the **abs** function is as

function "abs"(arg : **unresolved_sfixed**) **return unresolved_sfixed**

The **abs** function increases the size of the argument as

abs(sfixed(a downto b)) → **sfixed(a + 1 downto b)**

Example 8.17 Assume that the number x is of type **sfixed**(3 **downto** −3), let x be equals to −16, i.e., x can be represented by the binary string 1000.000. Then, $abs(x)$ produces +16 which can be represented by the binary string 01000.000, i.e., length of the integer part increase.

Example 8.18 The use of **abs** function is illustrated in PR 8.12.

PR 8.12 Program 8.12

```
library ieee_proposed;
use ieee_proposed.fixed_pkg.all;

entity fixedTest is
 port( num: in sfixed(3 downto -3);
        abs_num: out sfixed(4 downto -3) );
end fixedTest;

architecture Behavioral of fixedTest is

begin
 abs_num<=abs(num);
end Behavioral;
```

8.2 Floating Point Numbers

Floating point numbers enable high precision implementation. However, any hardware constructed using the floating point numbers is at least three times more complex than the hardware constructed using the fixed point numbers implementing the same algorithm or circuit. In addition, the circuits designed using the floating point numbers are much slower than the circuits constructed using the fixed point numbers. Considering these factors, although at the beginning, floating point numbers may be attractive a lot, however, from implementation point of view, they are not too much desirable.

Floating Point Numbers

Floating point number format is introduced in VHDL-2008. For this reason, during the project creation, VHDL-2008 language should be chosen for editing. Some of the VHDL development platforms may not support floating point number formats. Floating point numbers can be signed or unsigned.

To define a signed fixed or an unsigned floating point number, we use the syntax

signal/variable num: **float**(a **downto** b);

To be able to use the floating point number format in our VHDL code, we need to include the lines

library ieee_proposed;
use ieee_proposed.float_pkg.all;

at the header of the VHDL program.

The format of the 32-bit floating point number is as in

$$\underbrace{S}_{Sign}\ \underbrace{E_7 E_6 E_5 E_4 E_3 E_2 E_1 E_0}_{Exponent}\ \underbrace{F_{-1} F_{-2} F_{-3} \cdots F_{-23}}_{Fraction}$$

where S indicates the sign bit, E_i is used for exponent, and F_i is used for fractional part. The value of the floating point number is calculated using

$$Value = (-2S + 1)\left(2^{Exponent\ Value - Exponent\ Base}\right)(1 + Fraction\ Value) \qquad (8.1)$$

where *Exponent Base* is calculated as

$$Exponent\ Base = 2^{|E|-1} - 1 \qquad (8.2)$$

where $|E|$ indicates the number of bits used in exponent part. The smallest value of the exponent width and fraction width is 3, i.e., **float(3 downto -3)**, the smallest floating point number is a 7-bit number.

Example 8.19 If exponent contains 8 bits, then the exponent base is calculated as

$$Exponent\ Base = 2^{8-1} - 1 \rightarrow Exponent\ Base = 127.$$

Example 8.20 If exponent contains 12 bits, then the exponent base is calculated as

$$Exponent\ Base = 2^{12-1} - 1 \rightarrow Exponent\ Base = 2047.$$

The formula in (8.1) can also be written as

$$Value = (-2S + 1)\left(2^{Exponent\ Value - 2^{|E|-1} + 1}\right)(1 + Fraction\ Value). \qquad (8.3)$$

Example 8.21 To calculate the value of the x in

$$\textbf{signal x : float}(3\ \textbf{downto} - 3) := \text{"1011011"}$$

we first group the bits as

$$\underbrace{1}_{sign}\ \underbrace{011}_{exponent}\ \underbrace{011}_{fraction}$$

from which it is seen that

$$Exponent\ value = 3 \quad Fraction\ Value = 0.375.$$

And using (8.1), we get

$$Value = -\left(2^{3-3}\right)(1 + 0.375) \rightarrow Value = -1.375.$$

Example 8.22 The value of x in

$$\textbf{signal} \, x : \textbf{float}(3 \, \textbf{downto} - 4) := \text{``00010110''}$$

can be calculated as

$$\underbrace{0}_{sign} \; \underbrace{001}_{exponent} \; \underbrace{0110}_{fraction}$$

$$Exponent \, value = 1 \; Fraction \, Value = 0.375.$$

and using (8.1), we get

$$Value = (2^{1-3})(1 + 0.375) \rightarrow Value = 0.34375.$$

Example 8.23 Express 0.25 in the number format

$$\textbf{float}(3 \, \textbf{downto} - 4).$$

Solution 8.23 The number 0.25 can be written as

$$2^{-2}$$

and considering (8.2), we can express 2^{-2} as

$$\underbrace{0}_{sign} \; \underbrace{001}_{exponent} \; \underbrace{0000}_{fraction} .$$

In VHDL-2008, the following floating point data types are defined in the float packages as

$$\textbf{type float is array (integer range} < > \textbf{)of std\_logic}$$
$$\textbf{subtype} \, \text{float} \, 32 \, \textbf{is float}(8 \, \textbf{downto} - 23)$$
$$\textbf{subtype} \, \text{float} \, 64 \, \textbf{is float}(11 \, \textbf{downto} - 52)$$
$$\textbf{subtype} \, \text{float} \, 128 \, \textbf{is float}(15 \, \textbf{downto} - 112)$$

8.2.1 *Floating Point Type Conversion Functions*

For a floating point number, type conversion functions are available. These functions can be summarized as follows:

resize(float, exponent_width, fraction_width) $\rightarrow$ Changes the size of a floating point number.

to_slv(float) $\rightarrow$ The binary string representing a **floating** point number is interpreted as a **std_logic_vector**.

to_std_logic_vector(float) → alias of **to_slv**(float).

to_stdlogicvector(float) → alias of **to_slv**(float).

to_sulv(float) → The binary string representing a **floating** point number is interpreted as a **std_ulogic_vector**.

to_std_ulogic_vector(float) → same as **to_sulv**(float).

to_stdulogicvector(float) → same as **to_sulv**(float).

to_float(real number,exponent_width,fraction_width) → The input number is converted into a binary string representing a floating point number.

to_float(std_logic_vector,exponent_width,fraction_width) → The std_logic_vector is interpreted as a floating point number.

to_float(std_ulogic_vector, exponent_width, fraction_width) → The std_ulogic_vector is interpreted as a floating point number.

to_float(integer, exponent_width, fraction_width) → The binary string representing an integer is interpreted as a floating point number.

to_float(ufixed, exponent_width, fraction_width) → The binary string representing an unsigned fixed number is interpreted as a floating point number.

to_float(sfixed, exponent_width, fraction_width) → The binary string representing a signed fixed number is interpreted as a floating point number.

to_float(signed, exponent_width, fraction_width) → The binary string representing a signed number is interpreted as a floating point number.

to_float(unsigned, exponent_width, fraction_width) → The binary string representing an unsigned number is interpreted as a floating point number.

to_signed(float) → The binary string representing a **floating point** number is interpreted as a binary string representing a signed integer.

to_unsigned(float, ...) → The binary string representing a **floating point** number is interpreted as a binary string representing an unsigned integer.

to_integer(float) → The binary string representing a **floating point** number is interpreted as a binary string representing an integer.

to_sfixed(float,left_index, right_index) → The binary string representing a **floating point** number is interpreted as a binary string representing a signed fixed point number.

to_ufixed (float, left_index, right_index) → The binary string representing a **floating point** number is interpreted as a binary string representing an unsigned fixed point number.

Now let's give some examples to illustrate the type conversion functions.

Example 8.24 The function **to_float**() can be called as

$$\text{ret\_val} < = \textbf{to\_float}(\text{arg}, \text{exponent\_width}, \text{fraction\_width}).$$

and the function **to_slv**() can be used as

$$\text{std\_logic\_vector} < = \textbf{to\_slv}(\text{float}).$$

The use of the conversion functions **to_slv**() and **to_float**() is illustrated in PR 8.13.

PR 8.13 Program 8.13

```
library ieee_proposed;
use ieee_proposed.float_pkg.all;
library ieee;
use ieee.std_logic_1164.all;

entity floatTest is
  port( num: out std_logic_vector(7 downto 0) );
end floatTest;

architecture Behavioral of floatTest is
  signal x: float(3 downto -4);
  signal y: std_logic_vector(7 downto 0);

begin
  x<=to_float(0.25,3,4);
  y<=to_slv(x);
  num<=y;
end Behavioral;
```

Example 8.25 The program in PR 8.14 illustrates the use of

PR 8.14 Program 8.14

```
library ieee_proposed;
use ieee_proposed.fixed_pkg.all;
use ieee_proposed.float_pkg.all;

library ieee;
use ieee.std_logic_1164.all;

entity floatTest is
  port( num: out std_logic_vector(7 downto 0) );
end floatTest;

architecture Behavioral of floatTest is

  signal x1: float(3 downto -4);
  signal x2: float(4 downto -6);
  signal x3: float(4 downto -6);
  signal x4: float(4 downto -6);

  signal f1: ufixed(7 downto -3);
  signal f2: sfixed(7 downto -3);

  subtype ufixed_7_3 is ufixed(7 downto -3);
  subtype sfixed_7_3 is sfixed(7 downto -3);

begin
  x1<=to_float(12,3,4);
  x2<=resize(x1,4,6);

  f1<=ufixed_7_3'("01101011010");
  f2<=sfixed_7_3'("11101000010");

  x3<=to_float(f1,4,6);
  x3<=to_float(f2,4,6);

end Behavioral;
```

resize(float, exponent_width, fraction_width)

to_float(ufixed, exponent_width, fraction_width)

to_float(sfixed, exponent_width, fraction_width)

Example 8.26 The program in PR 8.15 illustrates the use of

PR 8.15 Program 8.15

```
library ieee_proposed;
use ieee_proposed.fixed_pkg.all;
use ieee_proposed.float_pkg.all;

library ieee;
use ieee.std_logic_1164.all;
use ieee.numeric_std.all;

entity floatTest is
end floatTest;

architecture Behavioral of floatTest is

  signal x1: float(3 downto -4);
  signal x2: float(3 downto -4);
  signal x3: float(3 downto -4);

  signal s1: signed(7 downto 0);
  signal u2: unsigned(7 downto 0);
  signal i3: integer range -128 to 127;

begin
  s1<=signed'("10001010");
  u2<=unsigned'("11010100");
  i3<=15;

  x1<=to_float(s1,3,4);
  x2<=to_float(u2,3,4);
  x3<=to_float(i3,3,4);

end Behavioral;
```

$$\text{to\_float}(\textbf{signed},\ \text{exponent\_width},\ \text{fraction\_width})$$

$$\text{to\_float}(\textbf{integer},\ \text{exponent\_width},\ \text{fraction\_width})$$

$$\text{to\_float}(\textbf{unsigned},\ \text{exponent\_width},\ \text{fraction\_width})$$

Example 8.27 Floating point numbers can be converted to **signed, unsigned** and **integer** numbers using the functions, to_**signed, to_unsigned, to_integer** as follows

$$\text{to\_signed}(\text{float})$$

$$\text{to\_unsigned}(\text{float})$$

$$\text{to\_integer}(\text{float})$$

Example 8.28 Floating point numbers can be converted to **signed fixed** and to **unsigned fixed** point numbers using the functions, **to_sfixed, to_ufixed** as follows

to_sfixed(float, left_index, right_index)

to_ufixed(float, left_index, right_index)

8.2.2 Operators for Floating Point Numbers

Arithmetic Operators

Arithmetic operators can also be used between floating point numbers. The arithmetic operators that can be used between two floating point numbers or between a floating number and an integer are

$$+, \quad -, \quad *, \quad /$$

In addition to these operators, we have the functions

rem, mod, abs

and these function can be used between two floating point numbers or between a floating number and an integer.

Logical operators

Since floating point numbers are represented by binary strings, then logical operations can be performed between floating point numbers. The logical operators that can be used for floating point numbers are:

and, nand, or, nor, xor, xnor, not

The shift functions that can operate on floating point numbers are

sll, srl, rol, ror, sla, sra

Comparison operators

The basic comparison operators that can be used between floating point numbers are:

$$=, \quad > =, \quad < =, \quad <, \quad >, \quad / =$$

Some functions for floating point numbers.

8.2.3   Copysign Function

The function Copysign with declaration

$$\textbf{function}\, \text{Copysign}(x, y\, : \, \textbf{float})\,\textbf{return float}$$

returns x with the sign of y.

Example 8.29 If the floating point numbers x and y have the values

$$x = \text{``\textbf{0}1011100''}\, y = \text{``\textbf{1}010''}$$

then Copysign(x,y) returns "**1**1011100".

Example 8.30 The use of Copysign function is illustrated in PR 8.16.

PR 8.16   Program 8.16

```
library ieee_proposed;
use ieee_proposed.float_pkg.all;

entity copySignTest is
end copySignTest;

architecture Behavioral of copySignTest is
  signal x, y, z: float(3 downto -4);

begin
  x<=to_float(0.25,3,4);
  y<=to_float(-0.25,3,4);
  z<=Copysign(x,y);
end Behavioral;
```

8.2.4   Scalb Function

The function Scalb with declarations

$$\textbf{function}\, \text{Scalb}(x\, : \, \textbf{float};\, n\, : \, \textbf{integer})\,\textbf{return float}$$
$$\textbf{function}\, \text{Scalb}(x\, : \, \textbf{float};\, n\, : \, \textbf{signed})\,\textbf{return float}$$

returns $x \times 2^n$.

Example 8.31 The use of Scalb function is illustrated in PR 8.17.

PR 8.17 Program 8.17

```
library ieee_proposed;
use ieee_proposed.float_pkg.all;
library ieee;
use ieee.numeric_std.all;

entity ScalbTest is
end ScalbTest;

architecture Behavioral of ScalbTest is
  signal x, y, z: float(3 downto -4);
  signal n: signed(3 downto 0);
begin
  x<=to_float(0.25,3,4);
  n<="1110";
  y<=Scalb(x,3);
  z<= Scalb(x,n);
end Behavioral;
```

8.2.5 Logb Function

The function Logb with declarations

$$\textbf{function}\,\text{Logb}(x : \textbf{float})\,\textbf{return integer};$$

$$\textbf{function}\,\text{Logb}(x : \textbf{float})\,\textbf{return signed};$$

returns the unbiased exponent of x.

Example 8.32 The use of Logb function is illustrated in PR 8.18

PR 8.18 Program 8.18

```
library ieee_proposed;
use ieee_proposed.float_pkg.all;

entity LogbTest is
end LogbTest;

architecture Behavioral of LogbTest is
  signal x: float(3 downto -4);
  signal n: integer;
begin
  x<=to_float(0.25,3,4);
  n<=Logb(x);
end Behavioral;
```

Problems

(1) Implement two complex numbers using fixed point numbers, and write a function that finds the product of two complex numbers.
(2) Repeat 1 using floating point numbers.
(3) Define a complex number vector using signed fixed point number format.
(4) Why do we use the resize function?
(5) What is the difference between an overloaded operator and overloaded function?
(6) Consider the numbers

$$\textbf{signal } a : \textbf{ufixed}(4 \textbf{ downto} - 3);$$
$$\textbf{signal } b : \textbf{ufixed}(12 \textbf{ downto} - 3);$$

Determine the sizes of integer and fractional parts of the numbers that result from the following operations

$$a \times b, \quad a + b, \quad a - b, \quad a/b, \quad b/a$$

(7) Explain the **scalb**() and **abs**() functions for **fixed** point numbers.
(8) Calculate the numerical value of 'a' in

$$\textbf{signal } a : \textbf{float}(6 \textbf{ downto} - 3) := \text{``1010101001''};$$

Bibliography

1. V.A. Pedroni, *Circuit Design and Simulation with VHDL*, 2nd edn. (The MIT Press, Cambridge, 2010)
2. P.P. Chu, *FPGA Prototyping by VHDL Examples: Xilinx Spartan-3 Version* (Wiley-Interscience, Hoboken, 2008)
3. A. Rushton, *VHDL for Logic Synthesis* (Wiley, Hoboken, 2011)
4. D.L. Perry, *VHDL: Programming By Example* (McGraw-Hill Education, New York, 2002)
5. P.J. Ashenden, *The Designer's Guide to VHDL*, 3rd edn. (Morgan Kaufmann, Burlington, 2008)
6. F. Tappero, B. Mealy, *Free Range VHDL*. (Free Range Factory, 2013)

© Springer Nature Singapore Pte Ltd. 2019   247
O. Gazi, *A Tutorial Introduction to VHDL Programming*,
https://doi.org/10.1007/978-981-13-2309-6

Index

© Springer Nature Singapore Pte Ltd. 2019
O. Gazi, *A Tutorial Introduction to VHDL Programming*,
https://doi.org/10.1007/978-981-13-2309-6

Printed in the United States
By Bookmasters